599

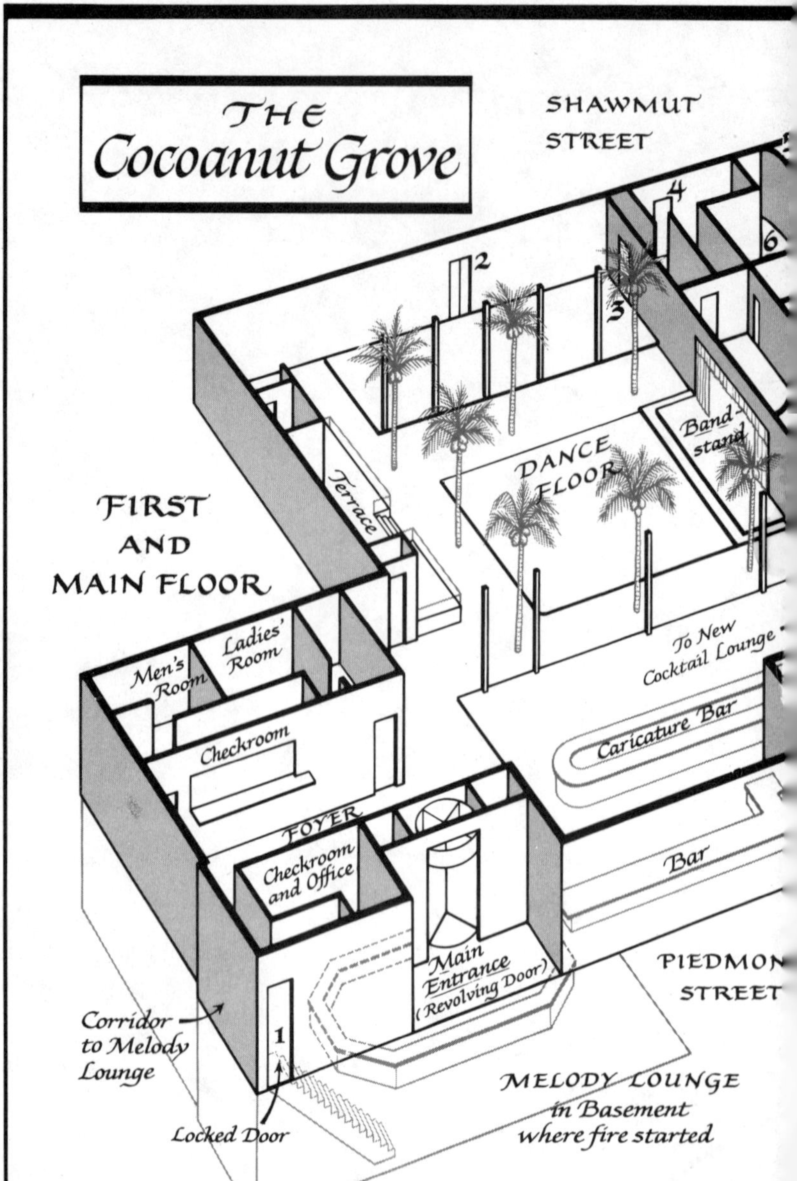
THE
Cocoanut Grove
SHAWMUT
STREET
FIRST
AND
MAIN FLOOR
2
3
4
6
DANCE
FLOOR
Band-
stand
Terrace
Men's
Room
Ladies'
Room
Checkroom
FOYER
Checkroom
and Office
To New
Cocktail Lounge
Caricature Bar
Bar
Main
Entrance
(Revolving Door)
PIEDMON
STREET
Corridor
to Melody
Lounge
1
Locked Door
MELODY LOUNGE
in Basement
where fire started
© 1984 A·Karl/J·Kemp

KEY

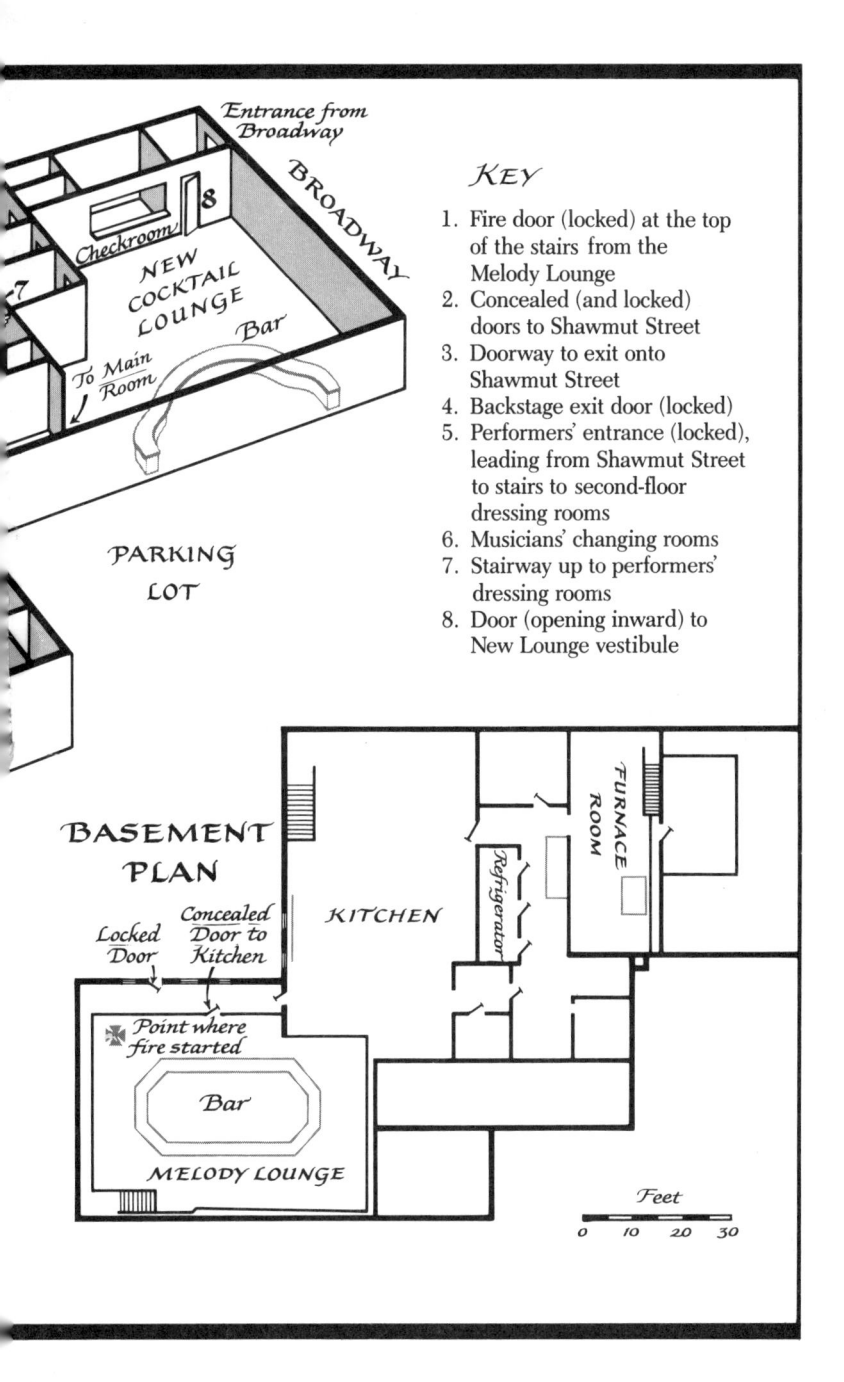

1. Fire door (locked) at the top of the stairs from the Melody Lounge
2. Concealed (and locked) doors to Shawmut Street
3. Doorway to exit onto Shawmut Street
4. Backstage exit door (locked)
5. Performers' entrance (locked), leading from Shawmut Street to stairs to second-floor dressing rooms
6. Musicians' changing rooms
7. Stairway up to performers' dressing rooms
8. Door (opening inward) to New Lounge vestibule

BOOKS BY EDWARD KEYES

Cocoanut Grove   *1984*
Double Dare   *1981*
The Michigan Murders   *1976*

COCOANUT GROVE

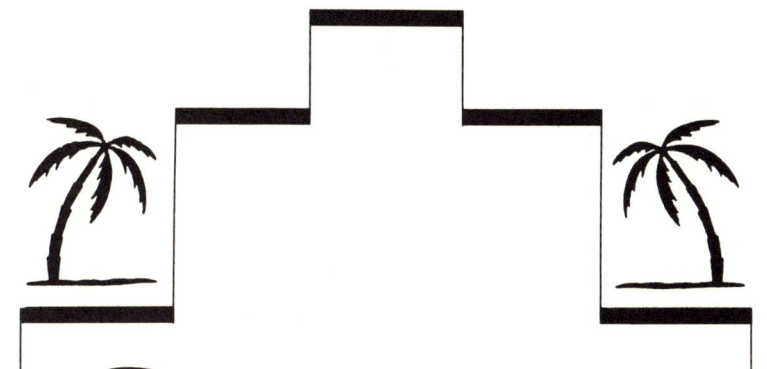

Cocoanut Grove

by

Edward Keyes

ATHENEUM • 1984 • NEW YORK

Library of Congress Cataloging in Publication Data

Keyes, Edward.
   Cocoanut Grove.

   1. Cocoanut Grove (Boston, Mass.)—Fire, 1942.
2. Boston (Mass.)—Fire, 1942. 3. Boston (Mass.)—
Nightclubs, music halls, etc. I. Title
F73.8.C62K49   1984   974.4'61042   83-45080
ISBN 0-689-11406-0

Copyright © 1984 by Edward Keyes
All rights reserved
Published simultaneously in Canada
by McClelland and Stewart Ltd.
Text set by Westchester Book Composition, Inc.,
Yorktown Heights, New York
Printed and bound by Fairfield Graphics,
Fairfield, Pennsylvania
Designed by Mary Ahern
First Edition

FOREWORD

It is always a problem for the author of a book such as this, having called upon so many to help bring it to life, to give due acknowledgment and thanks. On one hand, it is difficult and may be unseemly to specify any who contributed more than others; at the same time, merely to list each contributor equally can be unfair to some.

And in this case, such enumeration would be unwieldy (if not uneconomical) as well, whatever the intent. Thus I pray to be forgiven for saying here only how very grateful I am to *all* who aided my preparation of this work: the scores of individuals I was able to locate, in all parts of the United States, who so generously afforded me personal notes and recollections of the Cocoanut Grove and of its time; and the dozens of others who, having got wind of the undertaking, sought me out to volunteer additional valuable information and leads to yet more.

A number of public agencies and private institutions, on all levels of government and society, also were materially helpful. (Among all these, there *is* one individual who, I feel, should be singled out, for his kind forbearance and cooperation were of inestimable worth: Chief Andrew E. O'Brien of the Boston Fire Department.)

On a more personal plane, I must express both thanks and deep respects to Alfred Knopf, Jr., Chairman of the Board of Atheneum, out of whose keen intuition grew the idea to attempt this book; to Editor in Chief Tom Stewart, for his guidance and patience; to Editor Judy Kern, for her expert polishing of my not always elegant prose; and, hardly least, to special editorial mentor Fredrica S. Friedman, whose critical suggestions were invaluable.

There is no bibliography appended to the text, for the simple reason that I know of only one other book published on the subject. Nonetheless, that one—*Holocaust!* by Paul Benzaquin (Henry Holt & Co., 1959)—warrants special recognition for its thoroughness and countless informative insights.

Nor is there any index of names, places, sources. For this is not intended as a scholarly reference work—though I trust all who consult it may find it fruitful as an historical chronicle.

Cocoanut Grove is presented as a straightforward account, as vivid yet sensitive and truthful as I could tell it so long after the fact, of a time of great human travail that I can only hope brings an engrossing sense of immediacy to all who relive it—even now, as I write these words, forty-one years later.

Edward Keyes
New York, November 1983

COCOANUT GROVE

The greatest of faults is to be conscious of none.

Thomas Carlyle

PROLOGUE

It was a time in America that was fast nearing its end. Nothing would ever be quite the same again.

The great majority of Americans today, if they could be magically transported back to the 1930s and early 1940s, would surely find life in those times to be at least curious, perhaps even quaint.

Electric streetcars, or trolleys, with their ubiquitous tangle of cables overhead, were the customary mode of urban transportation. For overland trips of any length, people took coal-fed, steam-engined trains. Domestic commercial aviation had scarcely got off the ground (and planes at best were twin-propped, and had to refuel every few hundred miles); to go abroad, there were primarily ocean liners, or, for the adventurous, lumbering seaplanes called "flying boats."

Automobiles had only manual transmissions—with stick shifts usually on the floorboards—and practically all were dark colored; all had running boards along the side, and some smaller coupe models featured open-air rumble seats in the back.

Many telephones still were the old standup, earpiece on a hook type; for any number beyond a local call one had to dial Operator; party lines remained common in many areas. There were no postal Zip Codes; and in many cities mail was delivered twice a day, even at home.

More households had old-fashioned iceboxes—with great blocks of ice regularly brought to one's doorstep by truck or dray—than up-to-date electric refrigerators. The common heating fuel was coal. Milk (in quart bottles, unhomogenized, with a neck of pure cream at the top) also was delivered door-to-door, frequently by horse-drawn wagon. Frozen foods were unheard of: meals were either cooked from scratch or came out of cans. Kids were force-fed cod-liver oil with breakfast all winter and castor oil or milk of magnesia when sick. Doctors routinely made house calls. Hardly anybody had medical insurance.

There were no such things as credit cards; one paid with cash or on an "installment plan."

The main source of home entertainment—other than house parties, card games, or board games like Monopoly or Parcheesi—was radio. (Television, even to the handful then aware of its recent development, was little more than a futuristic dream.) All the national networks broadcast, Monday through Friday, from 10:00 A.M. to late afternoon, fifteen-minute dramas called "soap operas" (most were sponsored by familiar brands of household soap products), and from 5:00 to 6:00 P.M. there were adventure serials for younger listeners between school and supper (usually sponsored by cereals). Evenings, families sat around the radio to listen to regular weekly hour or half-hour dramatic presentations, comedy/variety programs, and game and quiz shows.

Local radio stations offered recorded music—big swing bands were the vogue, along with romantic (mostly male) crooners—as well as late-night live "remotes" from glamorous nightspots or ballrooms. The rhythms people most liked dancing to were the languorous two-step (or fox trot) and the bouncy lindy hop.

Nights out were given to dining, dancing, or the movies. Most movie houses showed double features, and some big-city palaces had live stage shows as well. Sex in films was almost always portrayed in terms of "romance"; lust was never more than implicit, and honor and goodness invariably won out. (Films were not individually "rated," but an industry-supported censorship group called the Hays Office had the power to blacklist any picture that did not adhere to a strict code of acceptable behavior. And the Catholic Legion of Decency regularly classified movies all the way from "Approved" for family viewing to "Objectionable in Part" or "Condemned.")

On the whole, traditional moral values prevailed: a girl risked a "fast" reputation if she kissed on the first date; virginity remained more the custom than the exception, and love and courtship were expected to result in marriage and procreation.

Men kept their hair closely barbered, and most were clean-shaven. The usual street attire for nonlaboring males, men and boys, was suits and ties, and hats—snap-brim fedoras for the more mature, flat "porkpies" or peaked caps for the snappier younger set. Many schoolboys still wore knickers. Adult women *always* wore dresses and hats (with seamed stockings), and some of the more ladylike affected white cotton gloves even in summer; younger females favored sweaters over pleated skirts, with anklets or shin-high "bobby sox" and two-toned saddle shoes or "penny" loafers. Female tennis players wore chaste skirts well below their knees. Women's swimsuits ("bathing suits") were one-piece. Shorts on either sex were reserved for resort areas.

It was very much a *white* America. "Minority groups" (a phrase one hardly ever used), meaning blacks—politely referred to as Negroes—had yet to gain any significant recognition and had virtually no place in the accepted social structure. (It was not until the United States was at war, for instance, that the Navy Department in 1942 even approved *recruitment* of Negro enlistees.)

Though America even then was widely depicted as a society rooted in violence—an image derived from highly publicized big-time gangsters like Al Capone and Dutch Shultz as well as colorful renegades like John Dillinger and Bonnie and

Clyde—at the broadest popular level the opposite was closer to the truth.

People walked the streets or rode public transportation at any hour of the day or night without fear. Doors commonly were left unlocked. Most shopkeepers felt no need to armor-plate their premises on closing. Policemen were legitimate authorities to be obeyed, even respected.

And even during the darkest days of the Depression, Americans as a whole remained trusting, peaceful, hopeful.

Despite a deep division in the country over whether or not it was America's moral obligation to join its European allies in their war for survival against fascist dictatorship, when finally America was forced into the war, its citizens unhesitatingly joined together in a remarkable fusion of solidarity and strength.

What it would be especially difficult to conceive for many people today is how ready, if not eager, the vast majority of Americans were to serve in that war. None wanted to die, few looked forward to killing or viewed the fighting in terms of heroics, but for most, it was something that needed to be done, a matter of personal honor.

In November of 1942, after almost a year at war, the mood in America was still urgent, hectic, yet beginning to show vibrant signs of renewed hope and confidence.

After close to a decade of crushing economic depression—seemingly chronic, and resistant even to the unorthodox, almost revolutionary measures enacted under President Franklin D. Roosevelt's "New Deal"—the destructive, humiliating aggression by imperialist Japan had come perilously close to burying an already dispirited nation.

But finally there were indications that the tide of endless defeat might be turning. The desperation of necessity had at last revitalized both industry and the people's will, and America was bristling with energy again. People were working; dollars flowed, and were being spent almost as though there might be no tomorrow—which was always a possibility, though few Americans wanted to concede it.

Hitler's Nazi Germany still held all of Europe in thrall, and the powerful Wehrmacht had poured across Russia to the very gates of Stalingrad and within sight of Moscow; the Japanese controlled all of Southeast Asia and were consolidating.

But there were small signs of changing fortune, and the air of America had become charged with returning optimism.

The national morale had been given a vital lift in April 1942, when bombers led by Major James Doolittle made a first retaliatory strike against the Japanese mainland, including Tokyo itself. In May, a reviving U.S. Pacific Fleet had battled a Japanese armada to a standstill in the Coral Sea off Australia. And then only a month later came America's first tangible victory of the disheartening war: the great naval battle of Midway, in which the enemy's inexorable advance across the Pacific finally had been stayed. In August, at last, the first U.S. offensive thrust: our Marines storming ashore on the key island of Guadalcanal in the Solomons—and holding on against fierce counterattack, and then slowly forcing the entrenched Japanese to retreat! Most recently, on the other side of the world, had come a combined U.S. and British invasion of Nazi-held North Africa and the first victories against the previously invincible German war machine.

And yet, at the same time another disquieting realization had penetrated the national consciousness: the United States was no longer beyond the reach of a determined foe. Early in 1942, a Japanese submarine had boldly surfaced off the coast of California and bombarded an oil refinery near Santa Barbara! Later in the year, another enemy sub had shelled the Oregon coast. Moreover, Japanese forces had occupied several of Alaska's strategic Aleutian Islands and were resisting all American efforts to dislodge them. In the Atlantic and Caribbean, too, wolfpacks of German U-boats continued to wreak havoc with U.S. and Allied shipping, often within sight of the American shoreline—once even brazenly torpedoing a freighter within the very mouth of the Mississippi near New Orleans. And on at least one occasion a sub had crept close enough to the beaches of Long Island to land a detachment of trained saboteurs (who, fortunately, had been detected almost at once and captured before they could do any damage).

Welcome optimism thus was tempered with grudging wariness. If enemy naval units could so easily invade our territorial waters, why not troopships, and aircraft as well? Newspapers daily published silhouettes and descriptions of both friendly and hostile warplanes. Along the seacoasts, volunteer watchers tirelessly scanned remote beaches and bleak skies, while in populated areas, civil-defense organizations were staging elaborate disaster drills to prepare the citizenry for attack.

Also in November 1942, the Selective Service System, in addition to processing millions of draftees eighteen years and older, began registering males up to age sixty-four. More than half a million American servicemen (and women) already were overseas, and at home, women increasingly filled wartime labor gaps as welders, electricians, mechanics, construction workers.

To help prolong and bolster production, "daylight saving time" was introduced—just "for the duration of the war," said President Roosevelt. The government banned further manufacture of radios and phonographs for civilian use. Anyone seeking to buy a fresh tube of toothpaste or shaving cream was required to turn in his used tube (to be recycled for war use). People began drives to collect aluminum foil, tin, and other scrap metals, and many started personal "victory gardens" to grow their own produce.

Rationing of priority commodities was instituted: sugar, meat, butter, coffee; and price ceilings were set on certain hard-to-obtain consumer products such as women's silk hosiery. (President Roosevelt even proposed a net income ceiling, for the duration, of $25,000 annually—thereby confirming for many of his opponents his true "socialist" leanings.) Gasoline rationing was to begin on December 1.

Thanksgiving turkeys, in short supply, were priced as high as 47 cents a pound. With beef also increasingly hard to get, horsemeat had been introduced experimentally at some markets.

In short, the war dominated every aspect of American life, in cities and hamlets across the country, as 1942 drew to a close.

So it was that November in Boston, the capital of Massachusetts. With some differences.

For one thing, the hubbub of war preparations was more apparent there than in most places. Though Boston was only a medium-sized city, despite its considerable reputation, its great port had made it a prime staging area for the fleets of military and merchant vessels continually steaming out across the Atlantic, and the city swarmed as few others with sailors, soldiers, and airmen.

For another, there was the quality of Boston itself.

While in most ordinary ways Boston differed little from any urban center, in other ways it was, and had long been, rather exceptional. For this city and its environs could (and did) boast of more culture, scholarship, advanced technology, and historical heritage than any other in the country; it was a city unique for its fine colleges and universities, superior medical and scientific research facilities, and authentic significance as a cornerstone of the American epic. It was sometimes called—and Bostonians felt no self-consciousness at the grandeur of the comparison—the "Athens of America."

On the other side of the coin, Boston had another unusual distinction. Founded and nurtured by Puritans, in time it had been infiltrated and finally overrun by masses of immigrants whom the haughty Yankees had tried harshly, and failed bitterly, to hold back—so that now, to the everlasting dismay of the founders' descendants, *they*, the Irish, Roman Catholics, were the majority, and they ran the city. Their patron saint was the one and only, "Himself," James Michael Curley: political power broker, former mayor, Massachusetts governor, U.S. Congressman, and devil incarnate to all Yankee Brahmins. Only that November, indeed, Curley—who as far back as 1903 had set a standard for his remarkable public career by being elected a city alderman even while in jail, convicted of official misconduct—at age sixty-eight had been returned to Congress for a second time.

Accordingly, contrary to the widely held notion that Boston tended to be rather a high-toned, stuffy preserve of bluebloods and blue laws, in fact it was, and had been for years, as lively, wide-open a city as any in the country outside of New York.

It was a sports-mad community and played host to a full complement of professional attractions—football, basketball, hockey, and *two* big-league baseball teams (the Red Sox and the Braves)—as well as intercollegiate competition, probably more than any other one city.

Only New York's Broadway had more legitimate theaters. And no place else had the sublime Boston Symphony, revered as one of the world's greatest orchestras, or its internationally beloved offspring, the Pops (*or* that aggregation's inimitable maestro, Arthur Fiedler).

Boston had an impressive array of distinguished hotels, many featuring elegant supper clubs with top-level entertainment and "name" dance orchestras, rivaling those of New York. And it also harbored a surprising number of first-rate nightclubs—more than a dozen in the downtown area, including the Latin Quarter, Steuben's, the Brown Derby, the Mayfair and, perhaps most famous of all, the Cocoanut Grove.

On an earthier level, burlesque flourished in Boston. The Globe, the Old Howard, and the Gaiety were but a few of the burlesque houses doing capacity business that November. And the notorious Scollay Square, not a "square" at all, but an area of several square blocks in the rundown north side teeming with lusty saloons and cabarets, always the mecca of

Boston's vast college population, also now swarmed with countless transient young servicemen.

It was Thanksgiving week, 1942, America's first Thanksgiving at war, and the Hub, as Boston was so aptly nicknamed, swirled with visitors and activity.

The big downtown department stores, notably Filene's and Jordan Marsh, were having big holiday sales. (Men's overcoats *or* suits, $15.50. Women's shoes, $1.99. Rayon stockings, 77 cents. Radios from $17.95.)

Theaters were busy. Among the first-run films then playing were *Across the Pacific*, starring Humphrey Bogart; *Desperate Journey*, with Ronald Reagan and Errol Flynn; John Wayne in *Flying Tigers*; *For Me and My Gal*, starring Judy Garland and Gene Kelly; Bob Hope and Bing Crosby's latest, *Road to Morocco*. And there was an imposing array of stage productions in town: Ruth Chatterton in Noel Coward's *Private Lives* at the Majestic; *Angel Street* at the Plymouth; *Junior Miss* at the Wilbur; a new comedy adapted from a Heywood Broun novel, *The Sun Field*, at the Colonial. On stage at the RKO Boston cinema was the new extravaganza, *George White's Scandals of 1943* ("60 Entertainers, 24 Beauties!!").

At the Opera House, the touring all-soldier revue *This Is the Army*, starring popular radio personality Ezra Stone, was playing its final week. The Casino Theater presented the Salmaggi Opera Company in *La Boheme, Lucia di Lammermoor, Pagliacci, The Barber of Seville*. And at Symphony Hall, Serge Koussevitsky led the Boston Symphony Orchestra before customarily packed houses.

The smart nightspots were thriving. Morton Downey was the attraction at the elegant Copley Plaza Hotel's Oval Room (where a complete Thanksgiving dinner, with show, cost a steep $3.50); Beatrice Kay headlined at the Latin Quarter; and at Boston's premier club, the Cocoanut Grove, the lineup included the dance team of Pierce & Rowland, Helene (and Her Violin), and the Carr Brothers, backed as usual by the local favorite, Mickey Alpert and his orchestra.

Every night place was booked solid. And to top it all off, there was a big college football game on Saturday the 28th that was the focus of considerable attention, bringing additional thousands of celebrants thronging into the city.

Not that the war was forgotten. Just the previous weekend, a mock disaster had been staged in Boston, testing for the first time the responses of some thirty thousand volunteer Civil Defense workers and the city at large to a supposed "blitz" by the German Luftwaffe. Air-raid sirens had begun whining at 3:00 P.M. Sunday; all activity throughout the city came to a halt, and pedestrians and motorists were directed to shelters. Over the space of an hour and a half, 169 "explosions" and "fires" were simulated at widespread points, demanding swift attention from fire, police, and emergency medical units, while a bemused but for the most part respectful populace looked on. The phantom attack was estimated to have produced some three hundred civilian "casualties," but overall the Civil Defense authorities were pleased with the response.

"Boston could handle an actual difficult situation," one organizer confidently proclaimed, "should it come tonight or tomorrow."

No one could have anticipated the following weekend.

chapter 1

The alarm rang at Fire Headquarters in the South End at 10:15 Saturday night—November 28, 1942.

It came from call box 1514, located at Carver and Stuart streets in the heart of the downtown entertainment district. It was an area of big hotels, theaters, and popular nightclubs, and was always crowded on weekends—and maybe especially so this busy Saturday night in Boston.

Firemen scrambled. Heavy motors roared up. Within two minutes, four engine and two ladder companies were careening out of the station in pursuit of one another, all klaxons blaring as they raced up Broadway.

In their second-floor dressing rooms, the entertainers at the Cocoanut Grove were killing time awaiting the late start of the evening's second floor show. They were to have gone on at 10:15, but, they'd been told, things were being held up while the club tried to squeeze a few more late arrivals into the already overflowing house downstairs.

The chorus and show girls, in full makeup and wearing the abbreviated Western costumes for their opening number, fidgeted restlessly. Some resumed an ongoing game of fantan, a few tried to read; others exchanged gripes about how diminished space on the dance floor, where extra tables were being set up, would hamper their routines. A couple fretted aloud over disruption of personal plans.

Chorus girl Jackie Maver, the lead dancer, had a heavy date with her sailor boyfriend right after the show; she didn't want to miss him—these days, you never knew when a guy might have to ship out. Her friend Pepper Russell tried to console her.

They heard the sirens and urgent rumble of fire trucks passing outside and went to the window. The dressing rooms faced out onto narrow Shawmut Street, behind the Grove, but were near enough to the corner to afford a partial view north on Broadway. The two young women saw the engines hurtle up Broadway and turn right on Stuart, then apparently stop and gather just around that corner.

"Damn!" exclaimed Pepper. "That's where the animal shelter is. I hope it's not there."

"Those poor little things..." Jackie murmured, her own annoyance forgotten for the moment.*

The fire companies had reached the location in less than five minutes. The men were more than normally tense: the previous Sunday, an explosion and fire in a building at Maverick Square in East Boston had killed six of their comrades, the worst Department tragedy in years; there'd been a solemn memorial service for them at the Cathedral of the Holy Cross just two days before Thanksgiving.

---

* Those familiar with present-day Boston, specifically in the vicinity of Park Square, may be puzzled by references to "the corner" of Shawmut Street and Broadway and a "view north" from there. Not many today would even know where to find Shawmut Street, which is no longer a "street" but hardly more than an alley about a half-block long whose former outlet to Broadway is blocked off by the wall of a multi-level parking garage; nor does Broadway—which used to continue north, beyond Stuart Street as far as Park Square—any longer extend farther than Piedmont Street, a block south of where it once met Shawmut. Both thoroughfares now end at the high-rise facades of the Howard Johnson's 57 Park Plaza Hotel complex, constructed almost exactly athwart the old location. What was in 1942 the corner of Shawmut and Broadway today would be about in the middle of the hotel's lobby.

14

What they found here, however, was hardly life-threatening: smoke pouring from the interior of a vacant automobile parked on Stuart just east of Broadway, in front of the Animal Rescue League building. A knot of passersby stood about. No one volunteered to say who had pulled the alarm a block away.

Relieved, though not without some sense of letdown, the firemen automatically smashed the windows of the locked vehicle—a nondescript, well-used black sedan with Massachusetts plates—and ripped out the smoldering front seat cushions, which they assumed had probably been ignited by a lit cigarette carelessly left behind by driver or passenger.

The small fire quickly extinguished, charred cushions stowed on a truck, the crew hauled in their gear and prepared to return to base. It was barely 10:20 P.M.

The premier happening in Boston this holiday weekend, the number-one subject of interest to area residents—and to much of surrounding New England—which above all else had contributed to the carnival atmosphere and swelled the city with greater than usual Saturday night crowds, was the college football game played that afternoon at Fenway Park: the traditional end-of-season battle for athletic leadership of Catholic Massachusetts between Boston College, the local favorite, and archrival Holy Cross of Worcester, forty miles to the west.

In an urban sprawl dominated by Catholics, Irish and otherwise, every face-off between these two Jesuit-run, all-male institutions was akin to class warfare, even in the calmest of times. Boston College was by far the larger school (a university, notwithstanding the "college" designation), with a more socially and economically representative student body; it was where Irish politicians and public servants sent their sons. Holy Cross, a trim little liberal-arts college, regarded itself as more selective, even elite—for the "lace-curtain" Irish, just a step below the Ivy League. Yet Holy Cross over the years had consistently fielded competitive athletic teams, some worthy of national recognition and often a match for BC: in the thirty-nine previous football meetings between the two since 1896, BC held only a slim 19-17 edge, with three ties.

This year, moreover, the contest shaped up as even more significant, especially to passionate and numerically superior BC partisans. The Eagles, as their athletes were called, were a powerhouse, undefeated and untied in nine games, ranked in some polls as the number-one college football team in the nation (led by their great halfback Mike Holovak, all eleven of their starting players had been nominated as All-Americans). Although it was not to be announced until after this final game, they had already been selected to play in the Sugar Bowl on New Year's Day, and the Sugar Bowl Committee had come up from New Orleans to present the honor formally—like a coronation—following the expected trampling of Holy Cross.

A gala victory party honoring the BC team and coaching staff—and by extension the whole sports-happy city of Boston—had been arranged to take place that evening at the Cocoanut Grove. A larger celebration for students and fans was set at the nearby Statler Hotel. Holy Cross was conceded no chance to upset these plans, not *this* year; the Crusaders, closing out a disappointingly lackluster season against a slate of opponents not nearly so formidable as BC, were listed on the betting line as 4-to-1 underdogs—and many thought those odds charitable.

The wet weather that had plagued the city earlier in the week had turned to dank cold by game time, as more than 41,000 lucky ticket-holders crowded into Fenway Park. (Among the more distinguished spectators were Mayor Maurice Tobin with a special guest, cowboy movie star Buck Jones, in town on the last leg of a cross-country bond-selling and morale-building personal-appearance tour.) Outside the old ballpark, radios across the city and over a wide area beyond were tuned to the game; activity throughout Boston proper slowed to a crawl, almost as if in reenactment of the air-raid drill the previous Sunday. The general atmosphere was electric with anticipation.

The game turned into a rout, sure enough—but hardly

as imagined. To the utter astonishment of nearly everyone, from the opening whistle the undermanned and underrated Crusaders outran, outpassed, and in every respect outplayed Boston College and, seemingly almost without breaking a sweat, mauled the mighty Eagles by an unbelievable 55 to 12! Fenway echoed to the exultant cheers of the minority Holy Cross rooters, while the mass of BC supporters gaped in stunned silence.

Before the end of the third quarter, the Sugar Bowl committee stole away from the park and made hasty plans for an early return to New Orleans, its invitation unbestowed.* Mayor Tobin and many others also left before the embarrassment was complete—with them Buck Jones, who was suffering, besides boredom, a bad cold and felt he should hoard his resources for that evening.

As for the Boston College Football Champions tribute at the Cocoanut Grove, it was quietly canceled by telephone from Fenway midway through the final period.†

The party mood buoying Boston all day was not so easily canceled, however. Thousands of residents and visitors had counted on whooping it up on Saturday, one way or another, and despite the somewhat dampening effect of the ignominious BC defeat, the city was thronged with spirited fun-seekers from late afternoon well into the night. The big Boston College supper dance at the Statler went on as planned, while Holy Cross had its own heady victory celebration at the Parker House. Theaters, restaurants and clubs all over town were doing turnaway business.

---

* Boston College was replaced in the Sugar Bowl by Tennessee against Tulsa University. BC did subsequently receive a bid to the Orange Bowl in Miami, where the Eagles lost to Alabama, 37–21.

† An ironical note to the debacle: The souvenir program for the game, made up weeks in advance, featured on its cover photographs of the BC cocaptains, Mike Holovak and center Fred Naumetz. The numbers they wore on their jerseys were 55 and 12!

At the Cocoanut Grove, the 7:30 dinner show was jammed. And by 9:00 P.M., as the early crowd was only beginning to thin, new arrivals were already milling about the rococo lobby and clustering outside on Piedmont Street. By all appearances, it looked like one of the club's biggest nights in years.

The Grove, built with California money and named after the Cocoanut Grove in Los Angeles' Ambassador Hotel (though the two were not related), had been a glamorous fixture in Boston since 1927. Both despite and because of its subsequent ownership by a flamboyant underworld figure and known bootlegger, who in the last years of Prohibition had turned it into a palatial speakeasy headlining top entertainment (but who himself was murdered in a rival club just before Repeal), it had become Boston's best known, most successful club, with an international reputation as well.

The Grove was located—along with its later competitors, the Latin Quarter and the Club Mayfair, both only blocks away—in a commercial section of the city south of the Common and bustling Park Square, in a warren of short, narrow lanes dense with low-lying structures of the garage or warehouse type. Boston was the regional distribution center for the motion-picture industry, and many of these utilitarian buildings were, or had been, used as so-called film exchanges. The Grove itself was housed in a typical single-story concrete structure, built in 1916, that had been both a film exchange depot (for Paramount Pictures) and a garage.

The marqueed entrance was on Piedmont Street, in the middle of a short block between Broadway and Church Street. Three stuccoed archways formed a portico, where revolving doors led into a plush red-carpeted lobby, or the Foyer as it was called. Ahead and to the right, beyond a checkroom, was the ornate ("swank" was the most frequent description) main dining and show room. It was a large square room whose most eye-catching feature was the seven tall, exotic imitation cocoanut palms with silver-tipped fronds that flanked three sides of the dance floor in the center. Beyond the dance floor, to the right rear as one walked in, was the elevated band stand sheltered under a striped canopy (underneath the stand was the movable stage which was rolled out over the dancing area for floor shows). Most tables, covered with white linen, were on the floor level around the stage, but along two walls were raised sections separated from the rest by wrought-iron balustrades—one, to the left of the entrance and up three steps, directly facing the orchestra and stage, was called The Terrace and was for VIP's and select patrons; the other was along the back wall, not as elevated but recessed under a tiled overhang.

The walls were off-white and rough-textured, hung here and there with pieces of colorful tapestry and heavy iron lanterns, and the ceiling was draped with a luxurious blue silken fabric, like royal tenting, so that the overall effect was pseudo-Moorish. (In pleasant weather, the fabric overhead would be removed and the "rolling roof" above opened to the night sky.) Overlooking this splendor, to the right of the entryway and a step up, was the elliptical, forty-eight-foot-long Caricature Bar, said to be Boston's longest, which featured hanging sketches of celebrated patrons.

Save for the bar, which had opened only after Repeal in 1933, and the recently installed rolling roof, this was the Cocoanut Grove as it had been from the beginning. By 1942, however, the club had grown into something more of a complex. In 1938, a section of the basement had been converted into an intimate hideaway called the Melody Lounge, featuring dim lighting and a discreet piano. And lately—just the past November 17, in fact—yet another extension of the Grove had made its debut: the New Cocoanut Grove Lounge, a sleek cocktail bar on the street level of a newly acquired adjoining building at the corner of Shawmut Street and Broadway.

And by 10:00 this Saturday night, all areas of the Grove were jam-packed—virtually every square inch of available space occupied, upstairs and down, with people still trying to get in. Management was doing everything it could to make a little more room....

* * *

It was about 10:22 when Pepper Russell heard the clatter of the fire engines returning to the station, unmistakable even without their sirens; it was only minutes after they'd first blared up Broadway. Glancing again out the dressing-room window, she was surprised to see that the apparatus had now pulled up by the corner of Shawmut Street. People appeared to be scurrying about. Outside the New Lounge? She could hear indistinct shouting. Now the firemen were jumping off their trucks, unfurling the heavy fire hoses—!

Pepper turned to the others in the room and cleared her throat: "Don't look now, but I think there may be trouble—"

Her uncertain warning was punctuated by a fierce pounding on the dressing room door. It flew open, and Charlie Mitchell, the club's assistant headwaiter, charged inside, panting, his eyes wild, face ashen and streaked with soot. "Get out, get out!" he rasped. "The whole damn place is on fire!"

chapter 2

By 10:00, waiters and busboys in the Grove's show room were having difficulty moving freely. All one hundred tables were long since filled, more had been maneuvered in among them, and still others were being set up on part of the dance floor. Every additional seat was occupied as soon as the chairs were put down. Customers stood elbow to elbow, and sometimes two or three deep, all around the long ellipse of the Caricature Bar. It was a noisy, high-spirited crowd, more so than on the average Saturday night.

At the roped-off entrance to the room, tuxedoed headwaiter Frank Balzerini and captain Leo Givonetti were striving with all possible tact to deflect the continuing surge of latecomers pleading, in some instances demanding, to be accommodated. And every so often they could not but submit to the insistence, or blandishment, of some well-recognized frequent patron with a party of guests who simply could *not* be disappointed. And so additional tables kept being sent inside.

Perhaps matters would have been handled with somewhat more panache and authority had the club's celebrated maître d'hôtel, Angelo Lippi, been at his customary station. But the suavely imperious "Count," as he was widely known, had been absent for two months, bedridden at home with crippling arthritis complicated by gout.

The presence of the Grove's owner of the past nine years, attorney Barnet ("Barney") Welansky, might also have made some difference, even though he was not the host type and normally kept himself well removed from the front of the house; but in any event Welansky, too, was ill, hospitalized since mid-November with a heart condition following a bout with pneumonia.

But perhaps, in the circumstances, the situation was simply beyond anyone's ability to cope. Several factors had contributed to the uncommon confusion. One, of course, was the special nature of the day in Boston. (Despite the Boston College defeat and cancellation of the planned victory party, the size of the crowd seemed even larger than expected.) Then there was the commitment Mickey Alpert and the orchestra had for a radio broadcast live from the Grove at 11:15 P.M., which made the show schedule all the tighter.

And finally, there had been the unexpected reservations, accepted quite late in the evening, for a group of important film-industry executives feting the veteran Western hero Buck Jones.

A party at the Grove had not been on the fifty-year-old actor's planned itinerary. Having come in by train from Washington, D.C., the night before, with only one full day scheduled in Boston, he'd been kept on the go almost every waking minute. First thing Saturday morning, he'd visited the Longwood Avenue Children's Hospital and made a triumphal round of the wards. Next, on a rented palomino and sporting his distinctive broad white Stetson, he'd ridden into the Boston Garden to the cheers of some ten thousand "Junior Commandos" and bona fide "Buck Jones Rangers." Then he'd attended a press luncheon in his honor given by area film exhibitors. From there, he'd continued on to Fenway Park as Mayor Tobin's guest. After suffering through three quarters with a worsening cold, Jones had left for a radio interview downtown. Next on the agenda had been a 7:00 P.M. cocktail reception in suburban Newton at the home of Monogram Pictures' district manager Herman Rifkin, attended by key regional motion picture executives and their wives, some up from New York, which Jones could not duck. Finally, at 9:30 he'd been scheduled to appear at the Buddies' Club on the Boston Common to meet some servicemen.

But by evening, aching with fatigue and the effects of his cold, Jones had asked Monogram's publicity representative to cancel the late stint at the Buddies' Club. He figured he would pay his obligatory respects to the film people, then slip back to the Statler and bed, to dream of entraining the next day for his long, relaxing return to California. Once his hosts at the party, however, had learned Jones was "free" for the balance of the evening, they'd insisted on treating him to a memorable sendoff with supper and a show at the Cocoanut Grove. And Jones, boxed in both socially and professionally, could only go along.

And so, a little after 9:30, the cowboy star and his entourage of thirty had descended on the Grove and been ushered through the jostling crowd in the Foyer to tables hastily reserved on the Terrace. Other customers turned to gawk, and the Jones party had scarcely been settled before a gaggle of autograph hunters and admirers pushed their way to the Terrace, thrusting menus to be autographed at the familiar square-jawed cowboy. All of which had only impeded the dining room's already overburdened staff the more; it was 10:00 before Jones and his group got drinks and could begin ordering supper.

By then the Latin relief band had finished its interim set and Mickey Alpert and his house musicians were reassembling and starting to tune up. Alpert, eying the waiters still struggling to manipulate new tables and chairs onto the gradually disappearing dance floor, was growing concerned for two reasons: one, the extra tables would greatly reduce the size of the stage and inhibit the movements of the dancing acts; and two, if the show were delayed too much longer it would run into the radio broadcast, and then everything would be screwed up. From the band platform, Alpert could see across the room out to the Foyer, and the confusion there did not appear to have slackened. Well, he thought, they're just going to *have* to cut it off pretty soon.

Billy Payne, the featured singer, came up to him. "They want you to go over and say hello to Buck Jones," Billy said, pointing out the film star.

Mickey glanced impatiently at his wristwatch. It was already 10:15. He sighed. "Might as well. But send somebody up to get the girls. We're running out of time here."

At the velvet rope, headwaiter Balzerini and captain Givonetti, no longer with a choice, had begun to hold firm: no more could be admitted to the club room; perhaps later, after the show. Would the gentleman and lady care to enjoy the Melody Lounge downstairs? Or sample the New Cocoanut Grove Lounge around the corner? Each quite cozy, with its own charming entertainment...

The Melody Lounge was directly below the main lobby, with the only access from within the club to the left of the arched Foyer, around a corner and down a flight of narrow stairs. The walls of the stairwell were of unfinished wood hung with fish netting, setting the tone for the vague sort of tropical exotica in prospect below. There, driftwood was set off with strips of rattan and bamboo, simulating coastal reeds, and in each corner rose a section of stylized palm tree with fronds overhanging the small tables and zebra-striped settees that hugged three sides of the room. Overhead, a deep-blue satin-like material billowed like a lowering night sky.

The only real illumination came from the indirect lighting over the enclosed octagonal bar that dominated the room. There were pinpoints of pale light among the artificial trees—which were electrically wired with 7½-watt bulbs interspersed among the greenery and cocoanut husks—but most of these were so artfully planted that their effect was little more than to cast a beguiling glow among the palms. This designed state of semidarkness was exactly what appealed to most who frequented the Melody Lounge; it was a place for romantics of all ages.

Enclosed behind the bar were three bartenders, two cashiers, and, on a revolving platform in the center, a cocktail pianist who played tinkly melodies. This night, however, the music was irrelevant, hardly audible behind the bar area itself. The "cozy" room could accommodate up to about a hundred customers without discomfort, but tonight there were perhaps twice that number—as many as four deep around the bar, shoulder to shoulder and knee to knee at every table. There were young lovers and middle-aged marrieds, in couples and larger parties, hopeful unattached GIs, and most of all boisterous football fans brandishing college banners, everyone trying to talk above everybody else. At the piano, in a flaming red dress, blond Goody Goodelle, a favorite on the Boston saloon circuit who had just been booked into the Grove, played on smilingly, mouthing lyrics if only to herself.

Behind his register at one side of the bar, cashier Daniel Weiss kept a sharp eye on the bartenders and the orders they were filling; with a mob scene like this, he knew, keeping tabs accurate could be quite a problem (as could the temptations for a little shortchanging). Weiss, a twenty-four-year-old fourth-year medical student at Boston University who had worked weekends at the Grove for the past three years, was the nephew of owner Barney Welansky. He never talked about it with other employees and so was not sure how many might be aware of his special relationship with management; nor did he know if any had guessed that his essential role there was as an undercover "spotter" for his uncle—moving from job to job, from area to area, upstairs, down, in the kitchen, watching for cheating or costly incompetence. Weiss didn't particularly care if it was known: family loyalty came first, and this job was helping to put him through med school. Besides, in another four months he would be a doctor, and his days at the Grove would be over.

It was about 10:15; Goody Goodelle had swung into the lusty wartime chanty "Bell Bottom Trousers," and a few at the bar nearest the piano had picked it up and begun singing along, with raised voices and waving hands cheerily urging others to join in... when the alert Weiss sensed rather than saw a stir in the corner of the room to the left of the stairs.

Moments earlier, he had noticed one of the white-jacketed bar boys conferring over the bar with head bartender John Bradley. The husky, neatly groomed youngster, whom Weiss had heard called Stanley—he was extra weekend help hired only a couple of weeks before—was asking Bradley what to do about a customer who had unscrewed one of the tiny light bulbs in a palm tree overhead: as faint a light as it had cast, without it that area was left in virtually total darkness. The boy had directed the barman's attention to the corner, where a young man and woman were nestling close under the tree. "Cramping his style," Bradley had drawled with a tight smile. "Never mind that. Go over and tell him, nicely, you've got to put the light back on—it's the fire law." Weiss had watched idly as Stanley made his way toward the corner, until he was distracted by a check to be rung up.

Now, moments later, from the corner of an eye, Weiss saw the boy return to the bar... and in the next instant the sudden flurry behind him, back in that corner. People in the immediate location were getting to their feet, some backing off, peering up, gesticulating uncertainly. (Yet others only a few seats away, occupied with themselves or concentrating on the singing around the piano, seemed oblivious of the small commotion.)

And then Weiss saw it: the spurt of blue light playing about the top of the palm tree where it met the lowered ceiling. It was so ephemeral Weiss had to blink to be sure his eyes were not playing tricks on him. But in the next instant the blue flicker became a ring of orange on the dark fabric above, outlining a black hole that at once seemed to grow deeper and wider. Then little jets of blue-orange flame broke out, up and down the imitation bamboo. Weiss's heart skipped.

"Get water, quick!" the cashier shouted to the bartenders. "There's a fire!"

"Jesus!" swore John Bradley. He grabbed a pitcher of water and scrambled out from under the bar, another of the bartenders following with a siphon bottle of seltzer, both yelling for waiters to bring more water. The bar boy, Stanley, was already back at the tree, swiping at the burning palm fronds with a bar towel.

The frantic splashing and spraying seemed to have little effect on the fire overhead. It would appear to go out in one spot only to flare anew in another. And now the crackling tongues of orange were licking out across the blue satin ceiling in a slowly widening circle.

The music had stopped—Goody Goodelle sat motionless at her piano, fingers frozen on the keyboard, staring over at the scene in the corner—but hardly anyone seemed to have noticed. The noise level remained constant. People on that side of the room all were standing, intent on the efforts of the employees to contain the blaze; yet few had made any concerted move to leave—as though immobilized by fascination or disbelief. At the far end of the bar, and at tables on the opposite side of the Melody Lounge, most remained seated, still gabbing and laughing, drinking and smoking, unaware of what was going on.

A few of those closest to the fire were, however, decidedly edgy.

Two couples sharing a table next to the one where the trouble began had witnessed the whole incongruous sequence of events: Maurice and Jean Levy and their companions Florence Zimmerman and Army Corporal Harold Goldenberg had noticed the amorous young man alongside unscrew the light bulb that had been annoying him and his date; they'd listened with some amusement to his protestations as the bar boy returned to explain haltingly why the light must be kept on; and they'd looked on as Stanley hauled himself up on the banquette and awkwardly fumbled among the palm fronds to find the socket. They'd watched—and Maurice Levy, for one, had felt the first sharp twinge of apprehension—as the boy, unable to see in the near darkness, struck a match close to the tree.

He'd located the socket and at once blown out the match, dropping it to the floor and stepping down on it for good measure, before replacing the loosened bulb. But in those few instants, Levy thought he'd seen something... what? A fleeting glint, a wisp flitting through the palm leaves?

The boy had apologized again to the displeasured couple and gone back toward the bar. But Levy had continued to stare up at the tree, and though what he thought he'd seen was no longer visible, he'd grasped his wife's hand and said to the others: "I think we should get away from here." They'd all smiled at his anxiety, but at his insistence Harold and Florence had risen with him and Jean and moved away...

And now there was no doubt about what Maurice had seen. A tantalizing flame nibbling at the fabric ceiling. Maurice felt dread mushrooming within him. They should get *out* of this place, *now*.

Another, younger group a few tables away also had risen. One of their foursome, Anne McArdle, an eighteen-year-old college sophomore, had excused herself just a minute or two earlier to go upstairs to the powder room; and now, with this little flareup in the tree, her companions were undecided whether to wait for her or pick her up on their way out. Actually, if they'd all been having a better time there might have been no question about staying—after all, it was only a flicker of a blaze, not all *that* serious, and it would be put out soon enough. But the fact was, the evening had been kind of a dud.

None of the three had met Anne before. Her escort, Jimmy Jenkins, a Harvard tennis captain, was a blind date arranged on short notice by a mutual friend (who happened to have been a longtime flame of Anne's, which didn't make breaking the ice any easier); and the other couple, Nathan Greer, also Harvard, and Kathleen O'Neil, were friends of Jimmy's. Anne had been described as saucy and fun, but she'd hardly showed it tonight, acting aloof, uninterested, and bored with the Grove.

So, standing, the three decided there wasn't much point

in lingering. Besides, they noted among themselves in making for the stairs, that little fire was beginning to look as though it might become trouble after all.

Another young woman had hurried upstairs moments after the fire broke out. Twenty-one-year-old Joyce Spector and her fiancé, Justin Morgan, were having a farewell dinner out, for Justin was off to the Army the following day. They'd planned to spend just a quiet evening together at Joyce's home in the West End; but at about 9:30 Justin had suggested they get some air, have a drink someplace, maybe dance a little. So they'd come to the Grove and, unable to get into the main club, settled for the Melody Lounge. Joyce had gussied herself up, wearing her brand new leopardskin coat with matching muff for which she'd scrimped $800 of her own money. They'd sat close, holding hands, murmuring through the din all around them; and the only other ones in the room they'd noticed was an older couple at the next table, whom they'd overheard telling a waiter it was their twenty-fifth wedding anniversary and who behaved as though they were still very much in love. Joyce and Justin had smiled wistfully at one another, wondering about their own future together.

But suddenly they saw the flash in the tree, and then in the ceiling practically over their heads, and Joyce had become scared. "Let's go, Justin," she'd pleaded. "Get the check, *please*. I'll grab my coat and meet you in the lobby." And she'd run up the stairs to the cloakroom.

Also toasting a wedding anniversay, their eleventh, were Donald and Mildred Jeffers. It had been a pleasant evening. They'd reprised their wedding supper, dining and dancing at the Hotel Statler, even treating themselves again to a bottle of champagne; this time, however, they'd gone out for a stroll after dinner, wandered hand in hand by the Cocoanut Grove, and decided to drop into the Melody Lounge for nightcaps.

Their quiet reminiscences had been jarringly disrupted, and Mildred in particular was seized by a sense of foreboding. Watching the employees' ineffective efforts to quell the still-small blaze, she gripped Donald's arm. "I think we ought to go!" His eyes on the growing circle of flame in the satin above, he said, "I think you're right. Come on." The two squeezed out from behind their table and began to edge through the wedges of people between them and the stairs.

Not even two minutes had passed since it began, yet those struggling with the fire sensed miserably that they were fast losing ground. The whole tree now was sheathed in flame. A waiter had rushed from the kitchen with an extinguisher, but the chemical stream was soon exhausted, without any appreciable effect. Despite all their efforts, the fire seemed to be spreading and growing in intensity.

Head bartender John Bradley and young Stanley were struggling to haul down the tree, which was now like a torch; if it could at least be isolated from the flammable ceiling fabric, they might make some headway. As they tugged at it with painfully blistered hands, fiery bits of bamboo and satin rained down over them, singeing their hair and faces. Then, at last, Bradley found some leverage and with a mighty yank tore the tree from its decorative mooring on the wall. With a shower of red hot cinder and spark, it toppled straight down on him, glancing off his shoulder, and he recoiled with a howl. Alongside, the bar boy jumped up to pull clear a section of the fabric ablaze over his head, but it disintegrated in his grasp, and he came down screaming, his arms full of flame.

It was too late; the fire, feeding on its own violent energy, was no longer controllable. With a fearsome *whoosh* it burst across the entire ceiling, devouring the fragile satin with incredible speed.

The spell gripping the onlookers was instantly shattered. Until then, they had seemed transfixed in contemplation of a phenomenon that simply should not—*could* not!—be happening. Now, in a flash of recognition, they snapped out of it as one. Bedlam was instantaneous.

There was a mass rush for the stairs, the only obvious exit, accompanied by a savage uproar of shouting and cursing, tables and chairs overturned or flung aside, glasses smashed, agonized cries of pain and despair as men and women were bowled over and trampled underfoot by the frenzied herd stampeding for safety.

Some made it to the stairs and up. But many were not quick enough. The flames, spewing thickening smoke and sizzling gases, roared about the windowless room almost like lightning, hungrily searching for an outlet. And within seconds a monstrous fireball raced for the narrow stairwell as though drawn to a chimney. It swooped upon those crushed together there, feverishly climbing one over the other, and exploded in their midst. Many were felled instantly by the superheated flames, while others, choking in the dense smoke and toxic fumes, were driven back down into the airless basement. In a moment, a pile of inert, charred bodies blocked the stairs. The escape route was cut off.

Some lucky ones did find a way out of the Melody Lounge that most in the desperate crowd were unaware of or had failed to notice. When the panic erupted, bartender John Bradley moved quickly to the wall beyond the bar and flung open the door to the club's kitchen—flush with the wall and decorated to conform with it, it was effectively concealed. Waving his arm frantically, Bradley shouted to all who could hear. The bar boy, Stanley, shepherded one group through, and Bradley collected about thirty more. Pianist Goody Goodelle followed with the female cashier. (In the melee, Goody had scrambled down from her platform, taken hold of the young woman, rooted in fright at her register—Goody knew she had a month-old baby at home and a soldier husband overseas—and guided her out under the bar.) They were the last to make it before the lights went out.

Others had been trying to get to that exit, gasping and stumbling through the smoke and debris, when the room was pitched into awful blackness. Falling, crawling, flailing about in blind frustration, none could find the door. One by one they subsided, scratching helplessly at floors and walls, emitting painful moans and dying wheezes.

And then—total silence. Nothing. Even the fire had disappeared, sweeping up and out the stairwell. Only a black hole remained, reeking with a miasma of hovering smoke, sickeningly sweet fumes, and the burned stench of death. Within a scant three minutes, all life in the Melody Lounge seemed to have been extinguished.

But at least one *had* survived, unhurt. Daniel Weiss, inside the bar, had hesitated when everyone first started running. Terrified as he was, he somehow felt bound, as Barney Welansky's nephew, to safeguard the bank, with which he'd been entrusted. He'd seen the other cashier leave her register; and, his mind in turmoil, was trying to decide what to do. Then he'd taken another look at the devastating scene before him—mass hysteria, bodies everywhere; some still sprawled across tables or slumped at the bar—and decided. He sprang from his chair and made for the gate underneath the bar, but before he could get there, the lights failed.

Dropping to his hands and knees in the blackness, he felt his way around inside the bar to the gate. He put his shoulder to it, but it scarcely budged; something was blocking it on the other side. Stabbed by panic, his primal instinct was to jump up and climb over... but then he thought of the smoke and gas, and those he'd seen propped up, asphyxiated, and he remembered that the only place there might still be safe air was close to the floor. Next to the jammed gate he knew there was a sink, and, keeping his head down—feeling the rasp in his own lungs now—he reached up and felt around. The sink was half full of wash water. He found a bar towel and soaked it with water. Then, covering his nose and mouth with the cool wet towel, he curled up on the floor, face down. And he lay there in the impenetrable, frightening darkness, breathing shallowly—but breathing!—and waited and listened.

Weiss realized all at once that the *only* sound now was his own breathing. There was not even the crackling of flames. Just a sudden eerie stillness. How long had it been? Probably only minutes, though to him it seemed forever in the darkness. What was happening?

And then he could wait no longer. He tried the gate again. Still no give. He could die there, like all the rest. He had to chance going over the bar. He resoaked the cloth, held it tight to his face, slowly rose up. Taking a last long breath, gripping the bar top, he lunged up and over.

He landed on a pile of bulky forms bunched on the other side and stumbled to his knees. *Bodies,* lifeless, in a heap blocking the bar gate! Swallowing hard to keep down the nausea, Weiss groped for the wall and the door. He found the doorframe and, lungs bursting for air, pushed through and fell out into the blessed chill of a smokeless passageway— only inches from where so many had just died for want of breath.

So far as he could tell, Daniel Weiss was the last living person to leave the Melody Lounge.

It was then just 10:20 P.M.

chapter 3

The dark passage into which Daniel Weiss had made his way from the Melody Lounge led, he knew, to the main kitchen and basement storage areas on the right, and through there to the service stairs to and from the dining room above. What he was not aware of, despite his several years' service at the Grove—as weekend help, he'd not taken the time to explore fully the labyrinthian bowels of the club—was that only a few strides to his left along that passage was a little used door that opened into an alley. He could not know, therefore, that only minutes earlier bartender John Bradley had shepherded a sizable group of fleeing customers, shaken but unharmed, to safety via that alleyway, followed by the bar boy, Stanley, with others. Weiss felt cold air in the passage coming from a vent somewhere, but his only aim was to get to the kitchen.

Feeling through the darkness, he found the door and lurched inside. There was some light in the spacious kitchen, from a single naked bulb dangling from the ceiling, and to his amazement Weiss found himself in the center of a group of wide-eyed people startled by his sudden appearance. There must have been two dozen, maybe more, some kitchen help but mostly customers and the majority of them women. Even in his own astonishment, Weiss could sense their extreme anxiety as they flung a babble of questions at him:

"Have you come from outside?"

"Is help on the way?"

"Can you get us *out* of here? Please—!"

Some plainly were on the brittle edge of hysteria. Weiss couldn't believe it. What were all these frightened people still doing down here? Feeling that somehow responsibility for their safety had been abruptly thrust upon him, he tried to calm them: "Now just take it easy. Of course you'll get out—we'll all get out."

Weiss spotted the club's food cashier, a matron in her fifties named Katherine Swett, whom he'd always thought of as "the Irish lady," standing firm by her register. He asked her: "Why haven't you gone upstairs?"

She stared back at him, shook her head, started to say something. But Weiss, impatient, did not wait for an answer. "Come on, follow me," he cried, starting for the rear of the kitchen and the stairs up to the main dining room.

Is help coming? How can we get out? Well, you're sure not helping yourselves staying huddled down in the kitchen, Weiss thought with annoyance as he bounded onto the stairs. He wondered again if people up in the club actually had any idea what had happened in the Lounge.

He was stopped in his tracks halfway up the stairs by a blast of heat from above. And then he heard once more the terrible sounds of screaming and the tumult of crashing furniture. Struck numb with horror, he realized it had never crossed his mind that the rest of the Grove—! Maybe he'd unconsciously refused to conceive it. But when the fire downstairs seemed to have spent itself so rapidly...

With a wrench he turned and ran back down to the kitchen.

None of them had followed him. And they only gawked now, their own worst fears reinforced. In a minute they would be blown into a million pitiful pieces.

Weiss remembered another flight of stairs from elsewhere in the basement that came up behind the bandstand above; there was a door there out onto Shawmut Street.

"The other stairs: have you tried them?" he demanded of the food cashier.

"Some have, from up there," she said, nodding toward the dining room. "They came running through—Mr. Alpert, some others." Her eyes darted questioningly to a door in the corner of the kitchen. "Nobody's come back..."

"Well, dammit, maybe they got out!" Weiss scolded. He looked at them all, so sheepish and scared, and regretted his impatience. "All right," he said, fighting to control the quaver in his own voice. "I'll go see. Everybody wait here. If it's clear, be ready to *move*."

He pushed through a swinging door into a dark cool room that contained the walk-in refirgeration lockers. Weiss made a mental note that they were commodious enough for a number of people if worse came to worst. Moving gingerly, arms outstretched before him, he came to a second door. This opened into another dark room and a sudden rush of hot air, and Weiss jumped back—before remembering it housed the club's heating plant. The furnace, going full blast, provided just enough light for him to make out the stairs rising along the back wall.

He inched up, one step at a time. There was an indistinct glow above. Then he could hear a jabber of voices, at first far off but soon closer, loud, urgent. Weiss was perhaps halfway up when from above there was a great splintering of wood and clanging of metal. Then a door at the top of the stairs burst open and a wave of cold night air washed down on him. In the same instant the cries above became whoops of release, as a horde of people clattered wildly through the shattered door and out into Shawmut Street. From below, Weiss could make out the silhouetted forms of firemen trying to push their way through the frenzied escapees into the burning club.

Breathing a fervent prayer of thanks, he turned back down the stairs, moving more swiftly now. Eyes accustomed to the darkness, he hurried through the furnace and refrigeration rooms and triumphantly threw open the door to the kitchen.

"Okay, people," he cried, "we've got a way out!"

They broke out in a confused gabble of excitement and anxiety. Some were afraid to chance leaving what seemed so far to be a safe haven. Weiss would hear none of that. "Quiet!" he ordered. "Now listen to me. Chances are *no* place inside the Grove will be safe for long. We must get *out!* You follow me, and we'll make it, I promise you." He had them all join hands and form a single file behind him.

Then Weiss noticed that Katherine Swett was not moving. He went to her and said, "What's the matter with you? Let's get out of here!"

The woman stood with one hand on her cash box, as though it were her last worldly possession. "I can't leave all this money just lying here. You go ahead. I'll be all right."

Weiss gaped at her, incredulous, and yet not without some understanding: hadn't he thought the same thing himself before he'd finally decided to flee the Melody Lounge? "Katherine," he pleaded, "this place could be coming down around our ears any minute. The hell with the cash—what is it, a few hundred dollars? We'll get it tomorrow. Come *on,* now!"

She shook her head doubtfully. "I shouldn't. It's my responsibility."

The customers were becoming more agitated. Two of the kitchen employees approached the cashier. "Katherine, you can't stay here! It'll be all right. Please!"

"I don't know," she said miserably.

The other two put their arms around her and gently urged her to the end of the ragged lineup. "Okay," Weiss called at last, "here we go." He pushed open the door to the refrigeration area. "Stay together, now!" he instructed, leading the way.

Their voices hushed apprehensively in the darkness. "Just a little farther," Weiss reassured them. "One more door ahead." He felt for the door to the furnace room. "Here we are. Almost home."

He swung open the door and stood with his back against it to guide them. A woman screamed.

"He's leading us into the fire!" another wailed.

Their collective control snapped like a rubber band. With cries of terror they broke ranks and rushed helter-skelter back toward the kitchen.

The horrified Weiss shouted after them, "Come back, come *back*! There's no danger here. We're almost out!"

But the last of them had already fled, heedless in their panic. He stumbled back through the refrigeration room after them. They huddled close together in the middle of the kitchen. "What are you scared of? That wasn't the fire—just the furnace! You've got to come with me!" he exhorted them. "It's suicide to stay down here! For God's sake, we can be out in two minutes!" He could smell the smoke in the air now; he could *see* it beginning to curl about the light bulb.

"*Please!*" beseeched Weiss. He appealed to the cashier, hovering once more over her bank, "Katherine—?"

"You go—leave us alone!" someone else croaked.

Weiss stared at them dismally, groping for something more to say. All he could think of was: "I'll send them down for you, if there's still time." Then, alone, he went through the swinging door again and out.

At the top of the stairs, he came upon people still straggling into Shawmut Street from the main room of the club. Most were singed, ragged, gasping, weeping; all were dazed, many close to shock, some barely conscious. Weiss grabbed hold of one ashen-faced man who seemed near collapse and half dragged him out to the sidewalk. Blocking their way, on his knees just across the threshold of the smashed doorway, was a black-clad priest praying over a fallen victim. Weiss, lugging an almost dead weight, roughly shoved the clergyman aside.

He lowered his burden to the ground and peered dumbly about. It was pandemonium: bodies tossed like rag dolls in the gutters; firemen, police officers, people running every which way shouting over the wailing and screams of the violated; fire trucks, ambulances, vehicles of all kinds; an incessant din of clanging and smashing and staccato hammering—like a riot scene, or a state of desperate siege. All this from one tiny flame down below? How long could it have been? He looked at his watch and couldn't believe what it read. He shook his wrist and held the watch to his ear, then looked again. It was not even 10:30. Less than a quarter of an hour.

Weiss was sent reeling by a hard shove from the side. "Keep movin', mister!" a grim-faced fireman growled, pushing past him. "We have to get in there."

Weiss sagged back against a wall of the Cocoanut Grove, suddenly spent and weak. Then, with a jolt, he straightened. "There are people downstairs!" he cried out, to no one in particular, to everyone.

The unsuspecting throng upstairs had had just two or three minutes' grace.

At about 10:15, even as the flames were beginning to smolder below, people had continued to mill in the Foyer—many persistently trying to wheedle their way into the club, some, having given up, leaving through the revolving doors, others standing around discussing what to do next, still more only arriving.

None would have yet had cause to notice the young woman hurrying from the far end of the Foyer toward the cloakroom. Joyce Spector, who had left the Melody Lounge to retrieve her new leopardskin coat, did not outwardly betray her concern or impart any sense of imminent danger.

Indeed, she even paused at the ladies' lounge off the Foyer to pass the word rather casually: sticking her head inside the outer door and finding the anteroom unoccupied, she called to the lavatory within: "Hey, anybody in there from downstairs? There's a little fire. Better stay up here." Hearing no response, she turned to go to the checkroom.

At that moment, the first terrified evacuees from below stormed around the corner and into the Foyer.

Those leading the headlong charge from the Melody

Lounge on reaching the top had hurled themselves at the emergency door onto Piedmont Street that was just to the left of the stairs. But it was bolted securely shut. That maddening delay, though it lasted only moments, caused the backup on the stairs that had turned civilized people into animals clawing for survival—and killed most of them.

Those who had made it to the top, finding the nearest outlet barred, rushed toward the only exit they knew—through the Foyer.

But so, too, did their pursuer, the fiery monster, driven by an explosive need for oxygen with which to replenish itself.

Among the first up were two of the college students, Nathan Greer and his date, Kathy O'Neil. (They thought Jimmy Jenkins was right behind, but he wasn't; he'd been mauled on the stairs by someone charging from behind and cast back down.) Running past the ladies' lounge, they stopped short. Kathy shoved open the door. "Anne! Are you there? Come out—*quick*! There's a fire!"

There was no chance to hear a response. In those few seconds the two were swamped by other escapees surging through the Foyer. (Among these were Maurice and Jean Levy. Their friends, Harold Goldenberg and Florence Zimmerman, were not with them.) Right behind hurtled the flames, bursting into the corridor as a huge, brilliant orange ball outlined in blue. Trailing black smoke, the ball raced the length of the vaulted Foyer in a flash, overtaking the fugitives. People's hair began to burn.

Joyce Spector, in a knot of people gathered at the checkroom, turned at the surprising commotion and saw the first spurt of flame and smoke bursting over the heads of those running toward her. She screamed, *"Justin!"* and she had but a single desperate moment to scan the confusion of faces before she was bowled over and trampled—still crying his name. (Justin Morgan, her Army-bound fiancé, who had waited to pay their bill, never reached the stairs and was trapped in the Melody Lounge.)

In the Foyer there was instant chaos as those already

there, immobilized at first, moved as one in a violent rush to the revolving doors.

The quickest and most determined of them got out.

To that moment—about 10:20 P.M.—the noisy over-flow crowd in the club dining room had no inkling of the nightmare that had just ravaged the room downstairs and was now streaking toward them.

It was a gay, expectant crowd, growing only mildly restless over the delay in the start of the floor show. The orchestra had been on the stage for over fifteen minutes, but nothing was happening, and here and there some younger patrons who wanted if nothing else some music to dance by had begun rhythmic clapping, more good-natured than insistent. It was an assembly representative of all ages and stations in life, most notable, perhaps, for its predominance of infrequent or even first-time nightclub visitors. Many had planned this Saturday night in anticipation of joining the gala salute to their great Boston College football team. But of course, in the wake of their shocking defeat, the team celebration had been canceled, and after a while the rest of the crowd had pretty much forgotten their disappointment and concentrated just on having a good time.

One popular member of the Boston College athletic community *had* come to the Grove, but few there would have recognized him or even known of him. Larry Kenney, athletic equipment manager for varsity sports, was a familiar and well-liked figure on the BC campus. He was fond of "his" kids and felt as bad as any of them about the humiliation by Holy Cross. But he was not about to let that spoil the big night out he and his wife had planned with their close friends. The two couples talked of the game over their drinks, and Larry told how during the previous week he'd kept putting up signs in the locker room warning the overconfident Eagles that Holy Cross might just give them a surprise or two. Then today, as the stunned BC players slumped by their lockers, some weeping openly in shame, Larry had chided them in his way that was bluff but never unkind: "Okay, so you lost. What are you guys, a bunch of babies? You know what? If we played 'em again tomorrow, we'd whip their butts! *You* know that, don't you? So forget it. It's not the end of the world!" Larry had hoped some of those kids might have squared their shoulders and turned up at the Grove anyway, just to show they still stood tall. But looking around at all the gaiety, he guessed he didn't much blame them for staying away. As for himself, though, it was strictly eat, drink, be merry, for tomorrow...well, tomorrow *would* be another day.

A few tables away sat a young doctor who a few years earlier had himself been an outstanding collegiate athlete in New England, a man whose name Larry certainly would have remembered. Gordon Bennett, from suburban Swampscott, had excelled in both hockey and football up at Dartmouth in the mid-30s, an indestructible tackle and co-captain of the 1936 Ivy League champion football squad (whose only defeat that season was, ironically enough, a 7–0 nonconference upset by Holy Cross). Bennett had gone on to Harvard Medical School and interned at Boston City Hospital, and now was a resident in gynecology at the Free Hospital for Women in Brookline. It was his first time ever in a club like this. He and his fiancée, high school sweetheart Edith Ledbetter, had attended the rout at Fenway, then agreed, despite Gordon's reluctance—for he was not much given to night life—to join their young married friends at the Grove and toast the others' first anniversary. Now, as much as he enjoyed the immediate company, Gordon was wishing they'd get on with the entertainment so he and Edith could leave at a decent hour.

Scattered among the roomful of celebrants there were many in military uniform—locals home on leave, others far from home, some having a last fling before shipping out.

Marine Captain Walter Goodpasture and his wife, Mary, were from Columbia, South Carolina. After extended sea duty, Captain Goodpasture's ship had put into Boston briefly to take on supplies before sailing again; and as it was the only port of call he could count on in the foreseeable future, Goodpasture had asked Mary to come up from Columbia and spend a few days with him—maybe their last ever together, who could say? They had been among the late arrivals at the club, but the clean-cut officer and his glowing young wife had little difficulty cajoling the guardians of the velvet rope into making a place for them inside. Now they sat close together at one of the tiny tables hurriedly set up on the dance floor and, oblivious to all around them, spoke reflectively of home and their love and the life together they dreamed of, the children they wanted, and of how long it might be before any of these things could come true.

At a table to the extreme left of the stage there was a small wedding party. John O'Neil and Claudia Nadeau had been married at 7:00 that evening at Our Lady of Pity Church in Cambridge, and with best man, John Doyle, and bridesmaid, Ann O'Neil, the bridegroom's sister, they'd come to the Grove for an intimate nuptial supper. The bridal couple— Claudia still proudly wearing her fragrant corsage of lilac blossoms and sweet peas, John, his carnation boutonniere—had posed beaming for the club photographer after pleading with her to take this one last shot. (She'd been about to leave, at about 9:45, to deliver a batch of negatives to her development lab down the street.) The photographer had said she would try to get the print back to them before the second show began; but now the show looked almost ready to go on and she still had not returned. The O'Neils hadn't really wanted to stay so late, they assured their best man and bridesmaid with coy smiles... but, what the heck, they *were* enjoying themselves. There would never be another night quite like this, and that picture surely would be a memento worth cherishing forever. They ordered more sparkling wine.

On the side of the room opposite the newlyweds, at a grouping of tables set back under the Mediterranean-style tiled overhang along the Shawmut Street wall of the club, sat a well-dressed party presided over by Boston's Civil Defense director, John Walsh. And on the Terrace, most conspicuous this night, were the congregation of film executives and their wives and friends feting the famous Buck Jones.

Most of the film group now were picking at their appetizer course, trading shop or family talk, and showing off pictures of their children to acquaintances they hadn't seen for some time. The guest of honor sat back wearily, in their midst yet somehow apart. Occasionally he would exchange words with his longtime associate Scotty Dunlap, a major producer of Western films and now Jones's personal manager, who had been along through all this taxing cross-country trek, or with Martin Sheridan, a local free-lance journalist retained by Monogram to help promote the star's Boston visit. But Jones was about smiled and small-talked out; his thoughts kept straying, as he confessed quietly to Dunlap, to the next day's train back home.

Marty Sheridan had about had it himself. He didn't much enjoy the role of press agent. It did provide useful supplementary income for a writer only in his twenties and not long married, trying to make ends meet on what he earned from irregular magazine assignments and stringing for out-of-town papers; but all he really wanted was to be a "real" newspaperman, and this sort of detour, even out of necessity, was less than personally gratifying. Nor was it easy money. He'd worked damned hard: weeks lining up a tight, productive schedule, following through on each step of the itinerary, sending out a welter of press releases, arranging interviews, appearances—even scurrying to find the right horse for Jones to ride into the Garden!—and then having to hold the star's hand every waking minute from the time he'd stepped off the train at the Back Bay station Friday night. All for one day's visit!

And then, some of those efforts had gone right down the drain when Jones fagged out and begged off some of his later commitments. Actually, though, Marty had been almost grateful for that, tired as he was himself; he'd hoped it would mean an earlier night than expected, a chance to surprise his wife Connie, who'd spent too many nights alone these past weeks.

But then the Monogram people had thought of this late fling at the Grove and insisted he bring his wife, and *that* was her suprise; she seemed to be loving it, so Marty guessed it had worked out all right. But still, even here he hadn't been able to relax, and enough was enough: he'd tried to catch the club's photographer for another picture of Jones, but had missed her; then the Monogram brass thought Jones should be introduced by the MC, with a spotlight on the Terrace, and Marty had sent word to Mickey Alpert—he would suggest a fanfare by the orchestra...

Marty was finishing his oyster cocktail, one eye on the bandstand watching for Alpert. He saw the orchestra leader start over, took a sip of his drink, and rose to greet him.

There was a raucous shout from somewhere behind Marty. Then another. It sounded like "Fight!" and seemed to come from near the entrance. He turned to look, as did others in the party, some half standing. Now he could see and hear quite a racket out in the lobby. A *hell* of a fight, he thought wryly—maybe he ought to go out and see if there was a story.

Marty looked back toward the stage. Mickey Alpert had stopped cold where Marty had last seen him, halfway across the dance floor. He stood rigid, looking startled and staring toward the entrance. As Marty squinted at him, Alpert's eyes moved up, over the heads of those seated, and his expression turned to something like horror. Marty followed his gaze, and he, too, froze: thick black tentacles of smoke were writhing overhead.

And then the shrieking was everywhere, and there was no mistaking the cry: *"Fire!"*

chapter 4

Up on the bandstand, the ten playing members of the Alpert orchestra had been dawdling, occupying or amusing themselves during the unusually long intermission with arranging and rearranging charts, tuning their instruments in individual riffs as musicians do before a set, gossiping idly among themselves. Electric as the Grove atmosphere was all this night, most of them were pros who had played many crowded rooms, and waiting was boring no matter where.

By the piano, violinist and arranger Bernie Fazioli, the *de facto* leader of the orchestra (Alpert was the boss, and fronted the group, but he was not at all musical), was fussing because pianist Moe Solomon had misplaced his sheets of a new tune, written especially for influential local newspaper columnist George Clark, that was to be introduced in this set, which meant Bernie had to hurriedly recopy the piano part.

At the rear of the platform, the youngest member of the band, Solomon's twenty-two-year-old cousin, Jack Lesberg, the bassist, absently fingered his strings, thinking of his date with the new cigarette girl, Bunny Leslie. Young as he was, Lesberg was in his second gig at the Grove, having returned only three weeks earlier from a tour with Muggsy Spanier's jazz group. It hadn't taken him long to spot the vivacious Bunny, and tonight was to be their first actual "date." Jack, however, had also taken a shine to another new addition to the club, the cute young singer/pianist in the New Lounge, Dorothy Myles, and he was thinking that he'd have to be careful to juggle the two interests artfully.

Lesberg's closest friend in the orchestra, despite a ten-year age difference, was saxophone/clarinetist Al Willet. Al, too, was dating one of the club's girls, dancer Pepper Russell. They had been pretty heavy for almost a year, really enjoyed one another; but Willet didn't know how much future there was in the relationship. Both were divorced and supporting children, she a small daughter, he a nine-year-old son, and he had doubts about whether a binding union—which was what Pepper seemed to be hoping for—was truly what he wanted so soon after his first unhappy experience. Noodling a few phrases on his sax, Al had his eyes on a large party to the left of the stand; there must have been seventy or seventy-five of them at a long row of tables, celebrating some couple's fiftieth wedding anniversary. All were elderly, yet brightly dressed and having a high old time together. Fifty years! That was something, all right. He tried to put himself in their place, but couldn't. Hell, if he remarried *tomorrow* he'd be in his eighties after fifty years! And he had no such plans.

Al turned and tooted a guttural vamp at Jack Lesberg behind him, and Jack grinned and returned a throbbing chord from his bass.

They both heard it at once: a shout from across the room that sounded like "Fight!" The two rolled their eyes at one another in exasperation.

It happened from time to time, at the Grove or anywhere else, especially on busy nights—some smart-ass elbows somebody, an argument starts, everybody overreacts. Al remembered the previous New Year's Eve, when there'd been a loud squabble at a table just to the right of the bandstand that almost turned into a gun battle. The whole orchestra had stopped playing and stood to see what was going on; and later Barney Welansky had chewed them out furiously, telling them never again to pay attention to things like that because it only made the situation worse. So Al Willet did not even turn around at once to catch the action.

But Jack Lesberg, facing the room from his more elevated position, could see that something was happening—not a fight that he could make out, but *something*. People were stirring, rising. The commotion seemed to be coming from the Foyer. Jack saw Mickey Alpert over near the Terrace. He'd stopped still, head cocked toward the entrance.

Bernie Fazioli turned to the musicians and said quickly: "Let's play something—"

None of them moved a muscle. All eyes had zeroed in on headwaiter Frank Balzerini, suddenly hurrying from the entrance past the bandstand toward the Shawmut Street side of the room. Balzerini looked intense. What—?

The impact hit like lightning: in one piercing moment, the chilling outburst of screams, people jumping up from tables—and the ominous black ball of smoke rolling through the archway, followed by a blast of flame that ripped across the ceiling over the Caricature Bar.

Lesberg sensed the musicians around him bolting from the stand, scurrying for the rear curtain, some with instruments in hand. Fazioli called in vain for them to stay put. Then only he and Lesberg and Al Willet remained.

Jack, supporting his cumbersome bass, stood rooted in place, mesmerized by the incredible scene before him. The noise all at once was earsplitting as, with the bellowing cries of the damned, people erupted every which way, wildly overturning chairs and tables, climbing over one another. There was a great confused surge toward the Foyer, bodies wedged tight, fighting for every inch. Then another phalanx, shoving, falling, scrambling, moved in a wave toward the Shawmut Street wall. Jack caught a glimspe of Frank Balzerini's tuxedo there. He and a group of other men were struggling feverishly to break open the doors that had been hidden behind a decorative screen. Gazing in disbelief at the raging bedlam, Jack spotted Mickey Alpert on the dance floor, arms upraised, shouting to the horde tumbling toward him... and then Mickey was swept along, lost among them. At the far end of the Terrace, Jack saw a blonde in a bright red dress reach the top of the stairs from the kitchen below. It was the new piano player in the Melody Lounge—holding on to another woman. And then they, too, disappeared in the heaving throng.

Al Willet, similarly paralyzed, had been watching something else: the progress of the fire over the Caricature Bar and around the room, advancing like a fuse on a stick of dynamite—sizzling rather than flaming across the old, dusty fabric of the ceiling, showering sparks and blazing bits of material down on the mob below. He saw men in panic climb atop the bar and leap wildly toward what they took to be the tall draped windows behind it on the Piedmont Street side— only to smash into solid glass and fall in heaps. They weren't windows at all but thick mirrors, frosted and blue-tinted and draped for the effect! Good lord, what a nightmare!

Then he saw the fire overhead streaking around toward the bandstand. Al broke out of his trance. He snatched up his sax and his clarinet and turned quickly to go.

Jack Lesberg was still frozen, he and Willet the only ones still on the stand. "Jack, we've got to get out!" his friend shouted.

At that instant, there was a rush of maddened patrons up onto the bandstand, and Lesberg came to life. He laid down his bass and, Willet alongside, scooted for the back curtain. There were steps behind there down to an exit onto Shawmut Street—

The lights went out. The screams rose.

Feeling their way in the dark, Lesberg and Willet got to the bottom of the steps and came upon a crush of people. The narrow doorway there, leading into a small vestibule and then the exit, was completely blocked. Al Maglitta, the drummer, was trying to get his bass drum and set of traps out. The drum was jammed in the doorway. Bernie Fazioli was with him, tugging and swearing, trying to help dislodge the bulky equipment.

"Let's go, for Christ's sake!" yelled Willet. "We're all gonna fry in here!"

"I can't, goddammit!" groaned Maglitta. "It's stuck tight!"

"Well, the hell with this! Jesus!" Willet and Lesberg heaved themselves against the drum, and with a scraping and tearing of metal and skin it tore free and bumped away ahead of them.

"You broke it!" cried Maglitta. Nobody answered him.

Willet sensed Lesberg turn from him, back up the steps toward the bandstand. "Jack! Where the hell are you going?"

"My bass—I should get my bass!"

"Are you crazy? We've got to get *out!*" Willet grabbed him by an arm. He threw down his own instruments. "They're not gonna do us any good if we don't."

The vestibule ahead, which also gave access directly to the dining room, was already jammed. The street door would not open. The shouting mob was pounding on it, ramming it, but it would not budge. More people, having clambered over the bandstand, piled down the steps, pressing in behind. Men choked out curses.

Lesberg and Willet and Fazioli were wedged together. Thick smoke was lowering upon them; bits of burning material flew in their faces, onto their shoulders and heads. They were trapped, no way to move. Unless that door was opened—

Al Willet gasped for breath. His throat rasped, he couldn't seem to draw enough breath. It was like inhaling a vacuum cleaner bag full of dust. He croaked to Lesberg alongside, "This looks like it, buddy..."

Jack said nothing. His lungs felt seared with each short breath. There was a sweetish odor that made his eyes swim; he was becoming light-headed. He felt he was going to pass out. He thought of his cousin Moe Solomon and his family. Parts of his life suddenly flashed clearly in his mind. Son of a gun! he thought. It really does happen—your life passing before you at the end. Then he couldn't think any more; he was floating, sinking, nothing beneath his feet. Into a black void.

* * *

The first startled shout of "fire!" had come from captain Leo Givonetti, in the Foyer, to headwaiter Frank Balzerini behind the velvet rope of the show room.

Balzerini had taken one horrified look at the monstrous ball of fire scooting along the ceiling, cried to Givonetti, "Get these people out!", meaning those clogging the lobby, then turned and hurried as fast as he could across the packed dining room to the Shawmut side to throw open the concealed emergency doors. He knew that if there was a mass rush to leave through the Foyer and main entrance it could be disastrous: not only was that whole area apt to be engulfed in flames at any second, but the revolving doors simply could not handle the hundreds of people all trying to leave at once; and if they ever jammed in such a crush—Lord, he didn't even want to think about *that*!

By training, the headwaiter's instinct was to avoid alarming the guests at all costs. Even so, he could see some beginning to stir apprehensively as he weaved among them. In front of the stage he signaled the musicians to the danger, knowing that if *they* got a headstart they could also open the other door behind the bandstand. The only other way out of the big room was through the corridor beyond the Caricature Bar that led into the New Lounge; but hardly anybody except club employees would even know about that door, much less that it was accessible.

When Balzerini reached the far side of the room, a group of patrons at tables under the eaves had already discarded the screen shielding the exit and were hurling themselves futilely at the locked double doors. At their head Balzerini recognized public safety chief John Walsh. The crowd around them was rapidly enlarging, and growing more agitated as the doors refused to give. The dapper headwaiter picked his way to the front. He knew these doors. One had a drop handle that had to be depressed sharply to spring both outward; sometimes it was balky.

But before he could get there, the wedge of grim-faced, perspiring men had reared back and made a last charge, and the doors burst open explosively. The people bunched together poured out into the street as through a break in a dam. Frank Balzerini was carried along with the tide and dumped on the sidewalk outside.

But that was not where he wanted to be, or felt he should be—a good headwaiter ought to be at his post, organizing, directing traffic, keeping order. Balzerini squared his shoulders and, despite the pleas of John Walsh and others, fought his way back inside the club.

Hardly more than a minute had gone by, but now it was pandemonium. The lights had gone out, heightening the panic. In the eerie illumination of flames overhead, people darted this way and that, like rats in a cage, slamming into one another, wailing and cursing through the flickering darkness; the din was unnerving, the shrill cries of fear and pain a counterpoint to the incessant deep-throated growl of the fire. The blaze had circled the top of the room with dizzying speed and was now working its way down the walls, the unbearable heat drawing closer and closer. People were gagging in the stifling smoke. People fell and bodies lay crumpled everywhere among the shambles.

Struggling against the unpredictable flow of disoriented fugitives—pausing here and there to direct one clutch or another of them to the doors now open to Shawmut Street—Balzerini made his way slowly back toward his station by the entrance to the dining room. Coming upon the elevated Terrace section, he thought to note how his VIP guests had fared. Peering across the wrought-iron railing, he straightened in shock: *everyone* in that exclusive preserve appeared to have been struck down! Still forms were strewn grotesquely, some entwined with upturned chairs, others piled atop one another on the floor.

Balzerini wrenched himself away—and turned straight into a fiery blast of smoke sweeping down upon him like a tornado. His tuxedo broke into flames. With a shriek he beat himself, dancing crazily; he reeled against a fallen table, spun off, and fell heavily; he tried to scramble back up, but fell again, and lay there, writhing... and in another moment he subsided, another blackened body, a slowly diminishing torch.

Frank Balzerini's worst fears about the potential for chaos in the main lobby had come to pass.

When the fireball from downstairs exploded in the Foyer, those already there, who were nearest the revolving doors, had the best chance to get out safely. On their heels, and charging into their midst, fighting them for escape, came those fleeing the Melody Lounge. And then, within only moments—as the fire leaped over the Caricature Bar and rushed through the main dining room—followed the next crush of escapees from inside. The Foyer became a teeming pen of terrified cattle.

Nathan Greer and Kathy O'Neil, two of the young foursome from downstairs, who had clung together hand in hand, were torn apart before they could get to the revolving doors. Kathy, straining desperately to hang on to him, fell away clutching his wristwatch, which had snapped off in her hand, as Greer was swept roughly out to the street. His instinct was to fight his way back in to her. But he could not buck the furious outgoing tide. He peered through the glass and called out in a misery of frustration, but there was no sign of her.

Maurice Levy and his wife, Jean, also were separated just before they reached the doors. Like Greer, Levy was pushed out into Piedmont Street, calling out pitifully to Jean as they were pried apart. His last memory of her was her bloodless face, eyes wide in fright, calling to him. Then she was lost in the churning mass.

Inevitably, under so much pressure, the revolving doors jammed—just as Balzerini had dreaded. Two men, hurriedly following their wives and a third lady in their party out to the street, got stuck halfway through. The women stood watching, aghast, just the other side of the glass, as the men struggled futilely to dislodge the doors. They would not budge.

And now the growing crowd behind them was pressing forward frantically. The raging fire and billowing smoke had begun to dip lower, drawn to the enticing draft created by the spinning doors before they'd seized. Some people were pricked by darts of flame and their hair or clothing caught fire. All were screaming. Some fell and were trampled where they lay.

At last, a mighty effort by those bunched in front snapped the snarled mechanism—the doors sprang loose, and bodies spun out as though catapulted, through the portico and across the sidewalk. In a few seconds perhaps two dozen people came out, gasping, stumbling, falling into the gutter, some burned, some bleeding, but alive. Some of the first tended to the last, and the fittest gathered themselves to help those still to come.

But now... no more came. The doors had jammed again. Those who had escaped pressed toward the entrance, staring through the unyielding glass at the horrified faces pressed against the other side. Suddenly a white-hot shaft of flame whipped out through the mangled doorframe and cracked panes, driving them back. The glass cubicle had turned into a pillar of fire!

Watching, helpless, from a few feet away, those outside could see that the doors were blocked now not by faulty mechanism but with bodies. All escape was cut off. They stood there wracked with frustration, watching people die. One by one their lifeless bodies buckled and collapsed, one atop another.

And the ones outside could do nothing to help. They screamed and wailed. Women fainted. Men cried unashamedly or threw up their guts.

chapter 5

Like every place in the Cocoanut Grove, the sleek New Lounge at the Broadway end had been filled to capacity, and more, much of the evening—all tables were continuously occupied, with customers at times two deep at the bar.

The club's new extension, on the ground floor of an otherwise unoccupied three-story tenement, had a motif unlike the rest of the Grove: it was decorated in pastels, with walls of dark leatherette brightened by an array of small oval mirrors and soft indirect lighting. It offered continuous musical entertainment nightly by performers who alternated from behind the curved, graceful bar. One set featured song stylist Maxine Coleman, imported from Hollywood, with Herb Lewis at the piano; the next, perky soloist Dorothy Myles. Though connected with the main club by the passageway, known and used only by employees, from the Caricature Bar, the New Lounge had been designed to be self-sustaining, and its only public access was an entrance off Broadway.

The stated capacity of the New Lounge was from eighty to a hundred. But the crowd this night had stayed well over that maxium. From a management standpoint its "debut" was turning out to be most auspicious; for although the room had been open almost two weeks, this was its first test under optimum conditions.

At about 10:20 P.M., the Grove's interim proprietor was

there counting the house. Jimmy Welansky—who managed or had interests in other local cafés, including the Rio Casino just a few blocks off—was standing in for his hospitalized older brother, Barney. Jimmy was there primarily for show: he was not really familiar with the day-to-day workings of the Grove, and left the running of the place to those who did know, the headwaiter, chef, bookkeeper, Mickey Alpert; he was just making himself visible as a representative for the Welansky stewardship. But he knew enough about restaurant operation to appreciate that by any standard the Grove was close to its peak that night. The New Lounge, indeed, might turn out to be the most satisfying of all to Barney, who had gone into Massachusetts General Hospital just before its grand opening. Jimmy could hardly wait to fill his brother in; if *this* volume of business didn't help speed his recovery, nothing could.

At this hour, Jimmy was performing one of his few official managerial functions, briefing the local police inspector on the setup of the new cocktail bar. It was the first visit to the New Lounge for Captain Joseph Buccigross, night commander of the South End district; among his responsibilities was personal supervision of security provisions at the various restaurants and cafés within his jurisdiction, but when this new room opened Buccigross had been away on vacation.

With them was Suffolk County Assistant District Attorney Garrett Byrne, who was not on official business. Byrne had dropped into the main club about 9:00, hoping to find some friends who he knew had planned to attend the BC victory celebration; he had run into quite a few, and after an hour or so was on his way out when he decided to look in on the new bar. Now he'd come upon Buccigross, who was in street clothes, and Jimmy Welansky, and the three of them had squeezed into a corner for a drink. Maxine Coleman and Herb Lewis were on, and it was hard to talk.

They were at the end of the bar farthest from the Broadway entrance, near an inconspicuous opening in the back wall that could have been the darkened entry to a service closet or pantry, but that in fact led into the passageway to and from the main club. Jimmy explained that his brother had not posted signs over it, didn't want customers to even know of it, lest wise guys come to use it as an escape route when trying to skip a big tab in the show room. Byrne and Buccigross had to exchange knowing smiles: that fox Barney Welansky—never missed a trick!

They'd been chatting only a minute or two when a woman hurried toward them from the passageway. She was a young, shapely blonde, wearing a flower in her hair and a dark sequined dress—a hostess, Buccigross thought, or one of the entertainers—but when she came into the full light of the New Lounge her face was pale and drawn with deep concern. She came straight up to the three men. "Mr. Welansky..." She gulped for breath. "There's a fire!"

Jimmy blanched. "*Fire?* Where?"

"In there." She gestured back toward the main club. "Everywhere..." Her voice broke.

The three gaped at her. Just then a puff of black smoke blew out of the passageway over their heads.

"Jesus!" swore Welansky.

Buccigross took him by the arm. "Let's see what we can do," he said, and the pair of them started into the darkened passageway. Byrne and the blonde stood watching them, wondering which way *they* should go.

The two hadn't advanced more than a few steps when around a corner of the passage stormed a flurry of people, headed by a white-jacketed waiter, rushing for the New Lounge.

Buccigross and Welansky, startled, stood dumbly in their path. Then the policeman raised his arms, exhorting: "Hold up! Don't panic!"

The restaurateur cried, "You want to start a riot?"

Their pleas were lost in incoherent shouts as the horde charged right through them into the lounge. Welansky was swept along in their momentum as though weightless, while Captain Buccigross, still commanding order, was sent sprawling.

And in their wake—even before these wild-eyed fugitives from the main club were out of the passageway—came the poisonous black superheated smoke bearing its insidious invisible gases, which upon contact with a new oxygen source immediately burst again into withering sheets of fire, as if from a flamethrower.

And with this, those crowded into the New Lounge, shocked momentarily by the remarkable intrusion of violence, themselves rose up in sudden frenzy and thrashed toward the street exit.

Garrett Bryne had already made it to the exit. The moment he'd spied that mob crashing through the passageway, and with an eye on the increasing influx of smoke, he'd headed across the room, assuming the young blond woman to be right behind. Near the door he'd stopped by a table occupied by two elderly couples and quietly but urgently advised them to leave at once—forget their belongings, just get out!—and for emphasis he had gone and opened the door that led into a vestibule and the outer doors onto Broadway. But as he'd pulled open that inner door—even as the room behind him erupted in flame—Byrne was shaken by a chilling realization. It wasn't scary enough that all these people would have to squeeze through the cramped vestibule to get to the street— the inside door did not push *out* into the vestibule, it opened *inward*! The place could be a death trap!

The assistant DA had no chance to dwell on the frightful prospect. The crowd behind him had now turned into a frantic mob, heaving ahead of it or hurling aside anything or anybody blocking its way. They crashed upon the doorway, and Byrne was shoved brutally through the vestibule into the street. Jimmy Welansky also was carried along.

So, luckily, was Captain Buccigross. Overrun and knocked down, he'd managed to scramble to his feet and, caught up in the surge, helpless to resist, was propelled across the lounge and flung out upon the sidewalk.

Few more got out, however, as the inner door, with people bunching up behind it, became an increasingly immovable barrier. Some of the men outside, one of them a tall young sailor, fought their way back into the vestibule and forced the door partly open to pull individuals clear—two of the first were hysterical, screaming women, both of whose clothing had been almost completely burned away. But it kept growing harder to get to them as the desperate mass inside backed up on themselves, only jamming the outlet more and more. It looked like an inferno in there. Some, in their desperation, began wildly attacking the glass-block windows with chairs or table legs, but only fragments of the tough glass broke out. The screams from inside were unbearable.

And then, miraculously, the fire engines were pulling up all around....

The fire companies returning to base from the routine automobile blaze a block away on Stuart Street, surprised as they were to be confronted suddenly by a fire in the Cocoanut Grove's new lounge, could have no conception of the size of the emergency. The men jumped off their trucks and busied themselves attending to the problem at hand, while the chief in charge dispatched a runner to sound an alarm from the nearest fire box.

That box, No. 1521, was located on Church Street, more than a block the other side of the Grove from Broadway. As the runner pounded along Shawmut Street past the back of the main club, there were no ominous indications of any problem within. The alarm was rung into fire headquarters at 10:23 P.M.

The firefighter then jogged around to Piedmont Street, planning to enter the Grove from the front and help initiate evacuation procedures; it was just possible they didn't even know yet that they had a fire in their new bar. Turning the corner, he stopped short in astonishment, and then horror. Shafts of flame and rolling black smoke were spewing from the club, over the heads of a crowd of battered people milling and shouting near the main entrance. Some lay collapsed in the street.

And at that instant there was a muffled roar like an explosion from the Shawmut side of the Grove. The firefighter spun around and raced back to the call box to sound another, more urgent alarm. That was at 10:24.

The men setting about battling the fire in the New Lounge were thunderstruck as the rear of the Grove seemed to blow out onto Shawmut Street, emitting smoke and flames and people tumbling outside. And they were further astounded as, not a minute later, additional equipment began clanging upon the scene: their own man could only just have rung in!* Nevertheless, assistance had come, not a moment too soon—for in that brief minute almost all the exterior of the nightclub, the whole square block of it, had become enveloped in roaring flames.

The only parts not yet afire were the two narrow, ramshackle buildings between the main club and Broadway. Were those buildings occupied?

The Grove entertainers—four female dancers or chorus girls, four chorus boys, four show girls, and, of course, the various specialty acts—were on the second and third floors of those old two- and three-story buildings. Barney Welansky had bought the properties and renovated the lofts while the New Lounge was under construction, and only two weeks before he had moved the entertainers up to the more spacious dressing and locker rooms, which for years had been down in the club basement, near the back stairs behind the stage.

In their new location, separated as it was from the rest of the club, the entertainers—still awaiting the call for the delayed second show—were the last to know of the havoc raging below.

---

* It would not be realized until later that in fact a first alarm had been sounded from box 1521 at approximately 10:20 P.M, evidently by some early, unidentified escapee, and thus responding companies were already on the way as the later alarms were received.

The assistant headwaiter, Charlie Mitchell, had come a long way to warn them. He'd been on the right side of the dining room near the Caricature Bar, with a clear view of the Foyer, when he heard the cry *"fire!"* Before he saw the threat he could hear it, a steady, intensifying *hisss*... and then he saw the smoke and jets of flame that turned into a fireball. For a second Mitchell froze, mesmerized like everyone else by that sizzling mass of fire floating uncannily toward him just beneath the vaulted ceiling of the Foyer. He snapped out of it only when he saw Frank Balzerini break from his post and hurry across the room toward the Shawmut Street side. Mitchell wheeled then and ran for the rear of the stage and the other exit out to Shawmut.

Mitchell was the first one at that door. He could not get it open; slamming with his shoulder did no good. In a moment others were swarming behind him, lunging feverishly against the unyielding barrier. One cried out in frustration, "Aren't there any exits in this goddam place?" Mitchell suddenly remembered another door—the one in the adjoining building that the show people used to get in and out of their new dressing rooms.

The show people! He'd forgotten about them. Did they have *any* idea what was going on down here?

Mitchell pushed clear of the growing crowd and picked his way into the musicians' changing room, which was empty, and out the other side into a dark corridor. A flight of stairs to the right led up to the dressing rooms; to his left was the closed street door. He tried it. Locked tight! He wasn't really surprised. The management probably had the door locked each evening after all the entertainers were inside to ensure their privacy and safety... and no doubt as another defense against deadbeats.

Smoke already was seeping into this passageway, oozing through the old walls as if they were porous. He could even feel the heat now. Mind in turmoil, Mitchell started for the stairs. He stopped as two men came sauntering down—the brother act, the acrobats.

"Do I smell smoke?" one said sniffing.

"You bet your ass!" cried Mitchell. "Come on, we've got to get this door open!"

The three hammered at it, but it was no use. Who had the goddam key? None of them knew. Then they were joined by another young fellow in skin-tight jeans—one of the chorus boys, a kid no more than eighteen. He looked scared. Mitchell sent him flying upstairs to warn the other male performers, with himself and the acrobats right behind. The smoke was getting thicker.

On the second floor the assistant headwaiter pounded on the door of the girls' dressing room, then threw it open. "Get out, get out! The whole damn place is on fire!"

Charlie Mitchell didn't take a lot of time describing how bad it was. He didn't have to—they could imagine the potential for catastrophe. Several of them had wandered downstairs between shows and seen the extraordinary crowds and the extra tables being wedged in around the dance floor. Jackie Maver had gone down to visit with friends who had come in just to see her, but she couldn't spot them from the wings and decided it was too crowded to try to find them; she would catch up with them after the late show, she told herself—for a few minutes, anyway, before she had to leave for her date. Now those missed friends were the first ones Jackie thought about, praying they'd been able to get out.

"Can we get down the stairs?" she asked Mitchell over the excitement buzzing through the room. Jackie (nicknamed MacGregor by the company for her Scottish ancestry and hardheaded practicality), though only twenty-five, was captain of the chorus line and used to taking charge.

"I wouldn't," he said. "It's filling with smoke and getting hot. And the street door is locked."

"What then? Do we fly?"

"We can get out on the next roof."

"Yeah, and then?"

"Climb down, drop down—whatever."

"Oh, great," Jackie groaned in chorus with the others. "Drop two stories! Talk about 'break a leg'!"

"It's better than staying here," Mitchell said, edging out the door. "That's where *I'm* going. You decide. Anybody wants to try it—" And then he was gone.

As the girls babbled at one another in dismay, Jackie went to the window and flung it open, a blast of icy air causing them all to shiver in their scanty cowgirl costumes. The street below was packed. To the right, at the corner of Broadway, everything was in confusion: a disorganized, shifting crowd spilling back into Shawmut Street, more fire trucks arriving and other vehicles clogging the intersection, people dashing this way and that. The night rang with sirens and horns, metallic clanging and urgent shouting voices. Then Jackie glanced down to her left and shuddered—though not from the cold: flames and black smoke licked out of the main club next door, knots of dazed people milled about, some wailing piteously; there were crumpled bodies sprawled on the sidewalk and in the gutter.

Jackie bit her lip. Back toward Broadway now she could make out helmeted and booted firemen beginning to emerge from the crowd and advancing along Shawmut, some lugging hoses. How long before they could get their ladders here?

The girls were crowding anxiously around her. Jackie faced them. "It looks like a 'wing and a prayer' whatever we do. Either go with him," indicating the departed Mitchell, "or wait up here for the firemen and hope, or try the stairs—and pray somebody gets that door open." She took a breath. "Me, I'm not about to jump, and I couldn't stand to wait. So that leaves the stairs. What do you say?"

"I'm with you, Mac," Pepper Russell spoke up. The others jabbered indecisively.

"Make up your minds," Jackie ordered. "We've got to go *now*."

Three more said they would chance the stairs. Three chose to wait for rescue.

Jackie said to those remaining behind, "Stay by the window. Let the firemen know we're coming down." Then she led the other four out into the dark hallway to the stairs, each with one hand over her mouth and nose and the other firmly on the shoulder of the girl ahead.

It was pitch-black on the stairs. The smoke was getting heavier the lower they got, but there were no flames yet. They inched down one step at a time, backs to the wall, holding tight to one another. The fumes began stinging their eyes and filtering through protective fingers, attacking nasal passages and throats. There was a constant muted undercurrent of sound, like a strong wind or rushing water, and it came to each of them that that was the sound of roaring fire. In the distance, they could hear an uneven, higher-pitched sound that wavered in and out like a short-wave radio, and that was unnerving because they recognized it as human screaming, and it would not stop. Without a light to follow or aim for, they might have been in a torturous descent to hell.

Henrietta "Pepper" Russell, fast behind Jackie Maver, her best friend at the Grove, felt she was about to die. Her initial fears and silent prayers had been for Al Willet. Mickey Alpert had tried to break them up, saying he didn't like his musicians to "mingle" with girls in the show. But for Pepper it was more than casual mingling. Though she was only twenty-one, while Al was in his thirties, she had more real fun with him than anybody she'd ever been with; and they had some important things in common: each had taken a hard fall in early marriage and they were both now raising children alone. She and Al fit together. He *mustn't* die—those had been her first thoughts. But now it was she, herself, on the brink, and her mind was filled with thoughts of her little girl—only two, suddenly to be alone and helpless, unable to understand where her mama had gone. Pepper could not stifle a loud sob, and the girl behind squeezed her shoulder reassuringly.

"Almost there," Jackie's disembodied voice came back hoarsely. Then there were no more steps, they were down, all wheezing and hacking, huddling together, blind and afraid in the impenetrable dark.

"Stand clear in there!" a harsh male voice echoed somewhere beyond them. And with a sudden abrasive splintering of wood and metal there was a gash of light ahead, and another, and then a rush of cold air sweeping away the smoke. Rough hands were passing them along to other rough hands, and they stumbled, gasping, weeping, into the wonderfully frigid night.

Pepper and Jackie stood numbly together, trembling from cold and shock, bewildered. The chaos, the din and carnage and agony, were disorienting, as though they had been thrust into a nightmare, another insane dimension. What had happened, *how* had it happened? The awareness of how close their own lives had come to forfeit welled up in them, and they clung together in wonder and relief.

They were only distantly aware of firemen brushing past them, hastily balancing ladders against the wall of the building they'd fled. But then an incongruous sound caught their attention—a man's long, appreciative wolf whistle!—and they gazed about to see people looking up at the building.

There, in their second-floor dressing room window, with only a wrap tossed over her skimpy costume, her bare legs prettily crossed over the sill, perched one of the show girls who'd stayed behind—looking for all the world in that one frozen moment like a Busby Berkeley creation—and rubber-coated men were scampering up the ladder to bring her and the others down.

Jackie and Pepper laughed until they had to cry. And then they kept crying.

chapter 6

It was like a war zone—a *real* war zone, and a grimmer test by far of Boston's capability to deal with catastrophe than even the Civil Defense planners of the previous Sunday's mock air attack had conceived.

The narrow streets and lanes around the Cocoanut Grove were not well suited to massive emergency and rescue efforts. By 10:30 P.M., as still more Boston fire companies had responded to the third alarm (the fourth, actually, counting the first that had, so fortunately as it had turned out, brought the equipment to the automobile blaze nearby), approaches to the Grove were so clogged with vehicles of all kinds that arriving apparatus could draw no closer than 150 yards, and those firemen had to run with their ladders and axes and haul their cumbersome hoses by foot the rest of the way.

The police had been mustering, too, speeding to the location from all over the city in radio cars, paddy wagons, commandeered taxis, trucks, and private cars, and on foot. Already crowd control was a problem, as throngs of awed spectators converged, threatening to hinder both incoming and outgoing emergency equipment. And there were others struggling to get to the scene: newspaper reporters, radio newsmen with their equipment, clergymen from parishes in the vicinity, military police and scores of other servicemen volunteering to help; and, of course, the official limousines convoying the city's shocked leadership—Mayor Tobin, Fire Commissioner William Reilly and Police Commissioner Joseph Timulty and all their attendant brass, Building Commissioner James Mooney, and the state fire marshal, the Massachusetts commissioner of public safety—down the line, top to bottom.

And over all, seeming to fill the night sky, sounded the interminable wail of the sirens. The sounds of alarm. Like war.

At first, firefighters trying to get into the blazing New Lounge had been driven back by the terrific heat buildup. They could get in as far as the vestibule, but the inner door had soon become blocked on the other side by the bodies of those who had failed to get out before they were overcome. One who had made it out safely and had rammed back in again and again, before the inner door jammed up, to pull one victim after another from that fiery entryway, was the tall, lanky young sailor. Finally unable to reach the door another time, he had staggered out aflame himself, his uniform virtually burned away, and collapsed on the sidewalk in a blackened heap. A small section of the tough opaque glass-block windows had been broken out from the inside, and protruding from it like a sculpted gargoyle was the death's head of a jowly middle-aged man who had tried to force his way through but had been impaled on the jagged glass.

Now a wedge of firemen wearing protective gear threw their bulk against the inner door and managed to open it enough to start dragging out the prostrate forms one by one while shafts of flame leaped out at them. Others smashed at the solid windows with heavy axes until hoses could be snaked inside.

The first streams of water shot into that dense, confined caldron turned to blistering fountains of steam. But surprisingly, it took just a few minutes of saturation to minimize the blaze. When the firefighters at last were able to penetrate the New Lounge—having extricated some three dozen bodies from behind and around the inner door—the concentration of heat inside was still intense; but they noted curiously that while there was much surface charring of furnishings, and though, of course, the place was a total shambles, comparatively little actually seemed to have been *burned*, that is, destroyed by the fire. Thus the primary destructive and lethal force must have been a combination of volatile, highly incendiary smoke and gases burning independent of the usual flammable materials. Still to be discovered, then, was the generative source of this phenomenon.

While some of these firemen felt their way gingerly through the white-hot and smoking passageway from the New Lounge into the darkened dining room, others on Shawmut Street, at the back side of the club, were exploring readier access.

The double doors broken open from inside the dining room, along with the locked door behind the stage that finally had been smashed through, appeared to have afforded outlets to the greatest number of patrons and employees, as well as a vent for the explosive pressure of flaming gases (hence the earlier sound of "explosion," which had quickly given rise to rumors in the street that a bomb had gone off in the Grove).

Even so, few had got out without damage; the number of the grievously, even hideously, injured was appalling, for the flames had swooped with them on their escape routes. Some had charged out with clothing and hair afire, literally human torches who threw themselves, shrieking, onto the icy sidewalk, there to stiffen into smoking hulks. Others had stumbled out appearing scarcely hurt, only to collapse suddenly after a few steps. People were sprawled everywhere in frightful disarray, many unconscious or dead—it was hard to distinguish—while others, weakly alive, crawled or lay about moaning in pain, choking for breath, suffering shock. These casualties were jumbled in the streets, impeding the rescuers; they had to be organized, moved out of the way. Fire officers swiftly began mobilizing the scores of civilian and military volunteers now hurrying upon the scene.

The fire continued to rage in the main club room, thriving in its spaciousness and on the infusion of air from outside; and out on Shawmut Street one could hear the piercing cries of those still inside who could no longer get through the wall of flame drawn to the openings. Torrents of water were directed through those openings, while firemen smashed at the three broad plate-glass windows that were painted over on the inside. Behind the shattered glass they discovered thin wooden partitions, and as they chopped through these it became evident that the windows on this side of the dining room—which could have been so easily broken for escape—had been concealed by a flimsy false wall decorated to match the rest of the interior.

Meanwhile, the first rescuers had penetrated the club proper by the backstage door. There, fire had been less damaging than smoke, and just inside the doorway, in the small vestibule and on the short flight of steps from the stage, they found a mound of prostrate bodies. Few of these were burned, but all were unconscious or dead. Among them were several men dressed in tuxedoes, one still clutching a battered violin.

All these were removed—laid out on the pavement in a lengthening row of victims now stretching all the way to Broadway—and then the firefighters, some in gas masks, ventured down a long, dark flight of stairs into the basement. Somebody had said people were still alive down there.

The silence below was incredible. There was no sign of flames, though considerable smoke deflected the beams of their electric torches, thickening as the men picked their way through the furnace room and the blackness of the refrigerator area and on to the kitchen. There, the smoke hung like a motionless fog, the hot stillness suffocating; but there was no fire. The place was littered with bodies, most of them women. Some were still conscious, pressing themselves to the concrete flooring with faces buried in towels or articles of clothing or just cupped hands; but others lay stiff and awkward, features bloated, expressions frozen.

One of the figures on the floor, a man, roused himself and looked up as the flashlight beam swept over him, and, holding a damp handkerchief to his nose and mouth, he struggled to his feet. He wore a rumpled, stained tuxedo, and a fireman recognized him as Billy Payne, the club's featured singer. Payne, voice raw and unsteady, directed the rescuers back to the refrigeration lockers: a handful of people had taken refuge in the walk-in units. It seemed like ages ago—he was worried they might have frozen by now!

Firemen hurried to the two large refrigerators in the adjoining room, each spacious enough to accommodate several average-sized adults. Pulling open the doors, they found four persons cowering in each locker, all ashen and shivering in temperatures only a few degrees above freezing. Two were women, one middle-aged, the other twentyish, and they were hysterical, refusing to come out; the fireman had to carry them. A man who looked to be in his mid-thirties was close to shock: glassy-eyed, he babbled numbly over and over, "My wife... where's my wife? We lost each other... Melody Lounge. My wife...?" Another man, dressed in a tuxedo, said he was one of the club musicians; outside, flapping his arms, he kept grumbling that he'd just been filling in for the regular trumpet player who was off taking his Army physical—man, he didn't need work *this* bad!

While these and the other kitchen survivors were hustled up the back stairs and out to safety, some firefighters remained below to complete the exploration of the basement. They found the exit to the alley, and the door into the Melody Lounge. It was pitch black in there, not a flicker of fire, only a bitter smell—and still as a tomb. Their torches picked out death wherever they flashed. A pile of bodies just inside the door to the kitchen. Several slumped over the bar, still almost upright! At tables. And the stairs—! God Almighty! Even the most senior of the firemen could hardly control his retching. If that guy's wife was in *here*...

The rescue teams up on Piedmont Street were also about

to encounter the horror of the Melody Lounge, but from the opposite perspective.

Nowhere was the confrontation with destruction more ghastly than at the front of the Cocoanut Grove, especially at the main entrance. The Foyer was still ablaze, and vicious jets of flame spewed through the broken but still jammed revolving doors. Helpless with rage and frustration, the firefighters could *see* the heap of corpses building only a few feet away, on the other side of the shattered glass panels. But until those doors could be unlocked, access was too limited to permit a mass assault. Individual firemen would crawl into the inferno to clutch an arm or leg and laboriously drag free one of the victims, but the flames and heat would force them back. Here, too, as on the Shawmut Street side, the row of dead and injured stretched out on the sidewalk was lengthening toward Broadway.

There were two other locked metal doors to the left of the beleaguered entrance. One was within the portico a few feet from the revolving doors, flush with the wall and without outside handle or knob. So far as could be determined, no one had escaped through it; either it was too securely locked, or it was not *seen* as an exit from the inside. It was difficult to get to, so close to the capriciously shooting flames, but under cover of a steady spray of water firemen with crowbars pried at it until finally it was sprung open. A new gust of pent-up fire and smoke drove them back; but hoses were turned into the opening, and in a few minutes crouching men carrying axes and hoses began to move inside. They found themselves in a small enclosed office, off the main Foyer, which appeared to double as a spare cloakroom. (A portable clothes rack was placed in front of the door to the street, so obviously it *was* unknown to patrons as an exit.) At last they had a foothold inside the burning club.

The other unadorned exterior door was some thirty feet along Piedmont to the left of the portico. This proved even harder to get open, as though double-locked on the inside.

At last they broke through—and stopped in their tracks, gaping at what they saw: a corridor filled wall to wall with bodies, one on top of another, some charred black, others unmarked but their faces horribly swollen, yellow or bright-pink pumpkins; and just inside, to the right of the door, a staircase leading to the Melody Lounge, the stairs also solid with inert, twisted forms. The firemen stared at the door they'd had to force open: it was supposed to be an escape hatch, a *fire* door, equipped with a "panic" lockbar that would secure it from the outside but open at a touch from inside—but it clearly had been double-bolted on the inside. How many were dead, trapped here and below, because that door had been inoperable?

Sick at heart and in their stomachs, the firemen had to lay down their tools and carry out each of the victims before they could continue inside.

Having once infiltrated the Grove at key points, the army of firefighters brought the flames rapidly under control. (The smoke and lingering fumes were something else, however, and those without gas masks could stay inside only for brief stretches of time.)

When it had been judged by all fire wisdom that no human being could possibly still be alive inside, tons of water in endless torrents were directed through every discovered and manufactured opening. More painted-over windows and fake interior walls were smashed in on the Piedmont side, as well as smaller ones just above ground level that gave access to the basement area—including the ostensibly windowless Melody Lounge. One after another, tall extension ladders went up against all sides of the Grove and adjoining buildings. Firemen climbed onto the club roof to chop out additional venting. (Someone thought of opening the famed rolling roof, and fire marshals struggled through the debris backstage to locate the electrical control box. But the unit had shorted out and the mechanism could not be activated.)

By shortly past 11:00 P.M., save for a few stubborn pockets

of flame, the fire was out. It had taken just about seven minutes to gut the entire Cocoanut Grove.

The enormous task remaining, which all but defied human resource, was disposition of the human wreckage.

The numbers of injured, dying, and dead were staggering beyond anyone's experience or conception. There were literally hundreds who had gotten out or been removed now filling the sidewalks and gutters around the entire block. Many were ambulatory, but more, it seemed, were not; and firemen probing the ruins were reporting corpses still inside that might total hundreds more. And then, the ashes and debris would have to be sifted and any remnants of personal belongings arduously collected and tagged to aid later in the soul-wrenching task of identifying those burned beyond recognition.

At 11:02, a final—in effect, the fifth—alarm was sent in to fire headquarters calling for additional manpower to join in the grisly mop-up.*

The paramount concern of the rescuers, once the fire was under some control, was getting the victims to medical facilities as speedily as possible. At first, while emergency resources across the city were being mobilized, the firemen and police commandeered any vehicle that happened to be in the vicinity—taxis, private cars, newspaper delivery vans and other commercial trucks, even some of the police radio cars responding—to transport early survivors to nearby hospitals. But soon all kinds of emergency equipment began to arrive

---

* In the end, the Boston Fire Department turned out no fewer than twenty-five engine companies, five ladder companies, one rescue company, and a water tower, plus assorted other vehicles including trucks equipped with spotlights and loudspeakers. In addition, as the extent of the disaster had become evident, fire departments from communities all around Boston, some as far as twenty-five miles away, voluntarily sent apparatus to assist; as the night wore on, the area south of Park Square, for blocks upon blocks in all directions, was filled solidly with fire equipment of just about every size and description.

on the scene—ambulances from the Red Cross and the Chelsea Naval Hospital, Army military police wagons, Civil Defense disaster vehicles (alerted by Director John Walsh as soon as he'd got out of the Grove)—along with scores of volunteers, doctors, nurses, Navy and Coast Guard corpsmen, and other servicemen, some hastily mustered in units by commanding officers at area installations to serve as extra rescue teams and litter-bearers.

Even as the operation thus grew more organized, however, several horrendous logistical problems became increasingly more apparent. One was the blockage of the narrow streets around the Grove, both by all the fire and rescue apparatus and by the growing throngs of spectators who kept pressing closer, severely impeding the movement of ambulances and other vital equipment. These streets had to be cleared to permit at least a steady flow to and from the disaster scene, and the police moreover had to set up fast, reliable routes to the various hospitals, blocking off or diverting all nonessential traffic.

Another difficult task, which fell mostly to the police, was that of crowd control. First there were the hundreds, quickly swelling to thousands, of horror-stricken and morbidly curious who flocked to the perimeters of the spectacular tragedy—passersby from the many bars and clubs in the vicinity, the theaters just letting out, the nearby hotels. Then, as the shocking news raced through Boston and environs, came the panicked relatives and friends of feared victims, urgently pushing, begging harried policemen to let them through. And inside the ring of disaster itself, there were the pitiable evacuees from the fire, many in escalating stages of hysteria, who, as they escaped or were brought out, milled dazedly about, searching bleakly or desperately among the battered, scorched forms stretched out on the sidewalks for loved ones or companions from whom they'd been separated. Understandable as was their anguish, they simply were getting more and more in the rescuers' way, and as gently as possible in the heartbreaking circumstances, they had to be led off—many in serious need themselves of medical attention.

Finally, the victims could not be removed fast enough, especially the dead and terribly injured, and these too began to hamper firefighting and rescue efforts; they clearly would have to be collected and placed somewhere out of the line of turbulent activity. After a consultation among fire and police officials, then, a multicar garage was requisitioned on Piedmont, just across from the Grove, and a warehouse around on Shawmut, and the bodies were hauled out of sight into what amounted to holding stations for the morgues.

By midnight, the only "victims" of the Cocoanut Grove remaining in clear view were the unclaimed automobiles, most remarkably undamaged, gathered in the club's small parking lot near the corner of Piedmont and Broadway. There were fifty-three of them.

Word had been passed—where it originated nobody was quite sure, though Jimmy Welansky was the assumed source—among the Grove entertainers and other employees who'd escaped, to gather at Welansky's Rio Casino, behind the Hotel Bradford several blocks away.

The glamorous-looking chorus and show girls particularly were grateful for some place to retreat; quite aside from the grief and shock and freezing temperature, it was difficult to put up with the reporters and photographers who clamored after them in their abbreviated costumes. In twos and threes they straggled away from the disaster area and walked numbly in the twenty-eight-degree cold to their refuge, oblivious to the stares and questions being hurled at them by the crowds.

Jackie Maver, captain of the chorus line, did not go direct to the Rio Casino. Still thinking of her date with Al "Whitey" Drolette, the Boston-based sailor who was her steady boyfriend, she detoured instead a block north to the Motor Mart garage across from the Statler, where she'd left her old small coupe. Then she hurried to the bar in the Avery Hotel near the Common that was their frequent rendezvous. Whitey was not there. The bartender, who knew them, told her the sailor had come in, ordered a drink, then heard the radio bulletins about the fire and dashed out—probably to look for her at the club.

Jackie hastened back to the Grove. But now she could not get even within view of the scene, as the police had succeeded in cordoning off a several-block area to all but emergency personnel and vehicles. Aware that Whitey would be frantic, not knowing if she was safe, yet despairing of finding him in the encircling mass of onlookers, Jackie at last made her way to the Rio Casino.

A restrained hush seemed to hang over the normally lively restaurant, even among the regular patrons who remained. The Grove survivors congregated at tables in the rear, close to the kitchen, gloomily nursing coffee and here and there a shot of liquor. Pepper Russell, from whom Jackie had separated soon after they got out, was one of several entertainers who had yet to appear. In subdued tones, occasionally with hollow irony, the group traded stories of their own or others' escape.

Charlie Mitchell, the assistant headwaiter who had alerted the chorus and show girls, once out on the roof had become alarmed by the heat of the tar beneath his feet and the smoke beginning to filter through from below. He'd climbed over the parapet and urged everyone to jump. Nobody else was ready to try it. To lead the way, Mitchell dropped two stories to the Grove parking lot. He landed hard and crumpled on the pavement in agony. Some sailors below, spotting the group stranded on the roof, called to the young women to jump into their arms; a couple of the more adventurous did, hurtling down screeching and with skirts flying, knocking their rescuers into heaps but getting up unhurt. The rest of the entertainers watched from above, still uncertain what to do. Then firemen with long ladders rushed up, threw them against both sides of the building, and brought them all down safely. No one else was hurt—only Charlie Mitchell, who was taken away in an ambulance, bones shattered in both legs.

They exchanged ominous rumors about people who hadn't shown up: Frank Balzerini, the headwaiter; Bunny Leslie, the pert cigarette girl everybody loved; Maxine Coleman, the singer in the New Lounge. Young Dorothy Myles, the pianist-singer who also played the New Lounge, was said to have been horribly burned and possibly dead. Mickey Alpert had gotten out; he'd been seen being led away, bleeding and in a stupor, some woman's white fur coat slung over his shoulders. There was no word of the popular Irish tenor Billy Payne. Several members of the orchestra were thought to have been trapped inside the club, but nobody knew who. Jackie Maver prayed wearily that Pepper's Al Willet was not among them.

Pepper Russell, once having regained her composure, refused to leave the fire scene. Tingling with dread and hope, shivering in her scanty costume, she stalked the rows of bodies along the sidewalks searching for the one she wanted more than anything not to find. Many of the victims were in gruesome condition, their skin blackened like leather, or crisp and discolored as parchment, or peeled away to reveal raw tissue and bone; some seemed hardly burned at all, their faces ruddy and only a little bloated, but frozen in death. She could not find Al. It was small reassurance: he could still be inside, or already carted away; or possibly he was among those burned beyond recognition. That was something she couldn't bear to think about. On the other hand, he might well be safe, maybe even as anxiously looking for her as she was for him. Pepper fingered the lucky fifty-cent piece on the silver chain about her neck and hoped.

On the Piedmont Street side of the Grove she encountered another costumed chorus girl, Connie Warren, being forcibly restrained by two policemen from clawing her way back into the still-burning club. Connie was frantic, close to hysteria: her husband, only recently hired as a cashier and this night assigned a register at the Caricature Bar, had not come out, she cried. She had to find him! Connie beseeched the officers. They held her back as gently as they could. No one but firemen could go back in there; it was suicide. Pepper took Connie in her arms and with a nod to the grateful cops walked her slowly away. Their men would be all right, she murmured consolingly to the sobbing young woman, wishing she were as sure of that as she tried to sound.

Red Cross workers took them in tow, threw blankets over their thinly clad shoulders, and led them off to the Bradford Hotel. Pepper vaguely remembered having been told to go on to the Rio Casino, but she was in no mood for glum post-mortems. The Bradford had opened a number of rooms to fugitives of the Grove, and she and Connie, numb with cold and both near exhaustion, weakly allowed themselves to be pampered with hot coffee, warm blankets and clean beds to stretch out on. Pepper thought for the first time that she ought to call home, to let her parents and her little girl know that she was all right. But she was so tired, she thought she would just rest awhile first.

She couldn't sleep. Al kept looming in her mind. She had to find Al—one way or another! Rousing herself, leaving Connie dozing fitfully, Pepper ignored all the well-meaning pleas to stay put and left the Bradford to return to the Grove. But now, like Jackie, she was unable to get back inside the police lines, no matter the urgency of her quest. Most if not all the victims by then had been removed anyway, she was assured. Where to? she demanded. Well, to practically every hospital in town—but most so far had been taken to Boston City Hospital, both injured and dead. The big Southern Mortuary was there....

Pepper walked several blocks before finding an empty cab heading back toward the Grove. When she told the driver City Hospital, he said he'd just come from there: he'd been cruising Park Square when the police had commandeered his taxi; they'd taken him down to Broadway and dumped a bunch of unconscious people into the rear and said to get them to City Hospital as fast as he could. He didn't know who was dead or who was alive; one or more was moaning, but he never looked around at them—he couldn't. It was the most awful thing he'd ever been through. When he'd gotten to the hospital emergency entrance, doctors and nurses had come running out and pulled the bodies out of his cab like slabs of meat. He couldn't help, he couldn't watch. He just got away from there as fast as he could. He really didn't know why he'd driven back to the fire. It just seemed the right thing to do. . . .

Pepper struggled through the chaos to the emergency reception desk. The sense of urgency pervading the hospital was dizzying: people dashing this way and that, people in white, policemen in blue, grey Red Cross people, soldiers, sailors, civilians. Bodies covered practically every inch of floor space in the emergency rotunda and stretched away down the corridors. Most of the first people she saw looked alive, though some just barely; many bore a reddish *M* on their foreheads, like laboratory exhibits. What did *M* mean? It was surreal. She could not have conceived so ghastly a scene.

Fighting faintness, she inquired at the desk if anyone named Willet had been brought in. A harried nurse scanned several sheets of roughly scrawled names and said no, not among the ones they'd been able to identify; some of the badly injured, and the dead, would take time to ID. Had she tried the mortuary? Many arriving DOA were now being sent straight there, to make room for those who could be helped.

Her heart icy with apprehension, Pepper picked her way among the casualties, many of them writhing and groaning or weeping, just as many ashen and still; she was only dimly aware that others around her were doing the same, bleak-faced men and women searching unfamiliar faces for one they wished desperately not to recognize.

She found herself in a long, wide, dingy corridor, with concrete floors and walls, that sloped downward below ground level. Orderlies hurried past wheeling hand wagons stacked with inert forms, and Pepper realized this must be the underground passageway to the mortuary across the street from the hospital.

She forced herself to continue. At last the corridor leveled off and she came upon an open area where the dead were being laid out in rows. Orderlies were tagging the corpses, and a couple of priests wandered among them, murmuring absolutions. The sight and the smell, and the realization, took Pepper's breath away, and she choked off a little cry of despair. She felt very near the breaking point.

One of the priests, a slim youngish man, came over to her and put a comforting arm around her shoulder. "Oh, father!" Pepper wailed, burying her face in his collar. There was a faint musty scent to him, like incense.

"Do you have someone here?" the priest asked softly.

"I don't know, I—"

There was a sudden shout from across the room. Pepper and the priest looked up.

"This one's alive!" an orderly cried excitedly. He was bending over one of the bodies in the center of the group. Others hurried to the spot. Eyes wide, Pepper edged closer. They were lifting the figure of a man whose head lolled from side to side; his eyes were closed, but his features were screwed up in pain. His face was streaked with soot, yet only his hair looked singed. He was wearing what might have been a tuxedo—

"Oh my God! It's Al!" shrieked Pepper. "It's *him!* He's alive!" She clung to the priest with all her remaining strength. "It's a miracle, father! A miracle!" She kept crying and laughing at the same time.

"Let's get him out of here!" called one of the orderlies.

chapter 7

On a Saturday night, and especially on a holiday weekend, the emergency department of any big-city hospital normally expects a high volume of traffic. Senior physicians and nursing supervisors may not be on hand, but extra interns and auxiliary nurses and even medical students and nurses in training are available to help the resident staff cope with the usual flow of random trouble.

But no hospital could possibly have been prepared for the emergency of Saturday night, November 28, 1942. And yet, quite by chance, Boston City Hospital happened that night to be as close to full readiness as if it had known what was coming.

Ironically, this was because of a holiday party taking place in the main nurses' residence, Vose House. Not only were many off-duty nurses and most of the resident students in attendance, but also dropping by to pay their courtesies were a number of the hospital's regular medical staff—including the administrator himself, Dr. James Manary—who would not otherwise have been there.

Dr. Manary had disengaged himself from the nurses' party about 10:00 P.M. Before retiring to his own residence on the grounds, he strolled through the eight-building complex, looking in on various wings and concluding his tour at the ever hectic accident receiving section on the ground floor, to the rear of the main building. There was nothing out of the ordinary—the usual rush of injured drunks and barbrawlers, auto-crash victims, emergency illnesses. Manary and the executive physician on duty, Dr. C. Winthrop O'Connell, spot-checked the several occupied examining and triage rooms and, satisfied, were about to bid one another good night when, just before 10:30, the initial alert was received at the emergency desk of a fire in the downtown entertainment district.

Minutes later a police car screeched up to the receiving entrance. In the back seat was an extensively burned female, only scraps of clothing still clinging to her horribly blackened flesh. A quick examination even before she was brought inside showed there was nothing that could be done for her, she was—perhaps mercifully—already dead. When the doctors heard it was the Cocoanut Grove, their hearts turned icy with foreboding.

They had someone telephone Vose House at once. The party was over.

Even as those doctors and nurses trooped back to the hospital—not without some annoyance, particularly among the younger student nurses whose rigorous training schedule afforded them little enough recreation—more victims were being brought in, and the brittle night air was increasingly filled with the shrill warnings of distant sirens. Disappointment was quickly replaced by apprehension and excitement as the staff began to sense the magnitude of the disaster.

Off-duty doctors were hurriedly paged at home or called in from wherever they had left word they could be reached— a black-tie performance at Symphony Hall, a suburban dinner party, even, in some cases, from vacation. Among those summoned was surgeon Charles Lund, in charge of City Hospital's new, federally funded research program for improved burn therapy. (With America's entry into World War II, the government had determined that burn damage, primarily to military personnel but also quite possibly among the civilian population, might be a critical problem; and it was recognized that methods of treating severe burns, whether from fire or chemicals, were still quite primitive when compared with the advances in other areas of modern medicine. Five hospitals had been awarded priority grants for such study by the presidentially appointed National Research Council—including City's crosstown "rival," Massachusetts General Hospital. The prestigious Mass General had been working intensively on the problem since early in 1942, while City had received its assignment only within the past several weeks.) Dr. Lund, who had followed the progress of Mass General's research with great interest and some envy, anxious now to close the gap, had left standing orders with City's night staff to call him, no matter the time, whenever the hospital received any third- or second-degree burn cases. By chance, an elderly man, badly burned when a cigarette set his bedclothes afire, had been brought in shortly after 10:00 P.M. that night, and Dr. Lund had been notified at home.

Lund reached the hospital within a half hour, five containers of precious blood plasma in his bag, and to his astonishment found himself walking smack into a far greater test than he'd bargained for.

The first victims from the Cocoanut Grove were delivered to City in ragtag lots—in taxis, police and fire marshals' vehicles, newspaper delivery trucks, Railway Express vans, along with a few ambulances and auxiliary station wagons hastily mobilized and dispatched by the Red Cross. Reception of these early arrivals was orderly and systematic, even if their growing numbers were dismaying to the hospital staff.

Under the direction of Dr. Manary, night supervisor O'Connell, and Lund, incoming victims were examined in the lobby of the accident room. Most were suffering various degrees of burns as well as exposure and shock; many had respiratory difficulties, indicating pulmonary damage. For pain, many were administered morphine (these were marked with a red *M* on forehead or chest). Soot and grime were quickly removed from exposed body surfaces, then those who required more extensive care were wheeled upstairs in the Dowling Surgical Building to one of two forty-bed wards only recently held open for Civil Defense casualties. There they received plasma and other infused therapy for shock, as well as oxygen where indicated; wounds were cleaned more thoroughly and serious burns sprayed with protective antibacterial chemicals—tannic acid, silver nitrate, deep violet "triple dye" solutions—to "tan" oozing burned areas against further loss of vital fluids.

What became more disturbing, however, with each new influx of victims, was the increasing number of those already dead. They were outnumbering survivors, it began to seem, by almost two to one. Some of the dead had sustained such enormous burn damage that they were not at first identifiable even as to sex; yet just as many seemed hardly burned at all. The only clue to the fate of these was a flushed complexion, even in death, indicating asphyxiation by carbon monoxide poisoning; there were others, moreover, who exhibited no outward signs at all of serious damage—they were just dead, by no immediately evident cause.

And the medical teams soon began to detect another trend that was even more mystifying: individuals would enter the hospital under their own power, seemingly suffering nothing more serious than dishevelment and nerves—no burns or clammy skin or trouble breathing. These were given superficial attention and shunted aside in favor of those obviously injured... and then, only minutes later, they would collapse without warning and, almost before anyone noticed, expire quietly within seconds. The lethal power of the Cocoanut Grove holocaust grew more awesome with every passing minute.

But the besieged doctors and nurses had less and less time to analyze or even to think about anything beyond the immediate situation. They kept coming—the walking wounded, the terribly injured, the dead. In less than an hour, the accident section lobby and all the triage rooms were packed wall to wall with Grove victims, alive and dead, stretched out side by side all the way back into the inner corridors of the hospital. The gathering stench of burned flesh and chemicals grew breathtaking, nauseating. The violet stain of triple dye was everywhere, blotting clothing, sheets, skin, implements.

And still they came—faster than they could be accommodated, much less properly examined. The rows of victims crying for attention soon extended outside into the courtyard and driveway and then onto Albany Street itself, and frantic medical personnel had to work their way out, trying to keep up with a crush that was fast becoming overwhelming. Soon, by awful necessity, examination of the incoming could be scarcely more than superficial: those showing signs of life were treated quickly, marked for medication, and maneuvered into the overflowing interior of the hospital; those who appeared beyond help were laid aside, to be piled on carts and removed to the Southern Mortuary across the street. Student nurses were assigned to stand grim watch over the dead, many of whom still wore furs and jewelry and other valuables. There were already morbid reports from the fire scene that some of the stricken had been cruelly looted.

There would be no way of telling how many of those "corpses" might yet have had the faintest breath of life (like the saxophonist Al Willet, retrieved from the mortuary). The daunting onslaught no longer allowed for infallible diagnosis; the worst of them simply had to be passed over for those with the most apparent chances for survival. Only later would doctors and nurses have time to agonize over the unanswerable question: who, if any, had died who might have been saved if things had been a little less chaotic that night.

Those brought in earliest were the more fortunate, for they received the fullest attention. One of these was a tall, lanky young man most of whose body, except for his face, was charred black as leather; only shreds of his clothing remained, fused to what little was left of his skin. It appeared to be the dark material of a uniform, and indeed whoever delivered him from the Grove had told somebody he was a sailor who had performed heroically, having gone back into the fire time and again to rescue other trapped patrons. His dog tags, warped from the intense heat, identified him as Clifford Johnson, age twenty-one, U.S. Coast Guard. He was very close to death—in fact, by all medical precedent he should already have been dead.

None of the staff had ever seen any living person burned half as badly as Clifford Johnson. He had sustained maximum third-degree burns over more than 50 percent of his body; in some places the layers of tissue were destroyed all the way to the bone; and another 25 percent of his flesh was painfully blistered by second-degree burns. It was a medical rarity at the time for any patient to survive third-degree burns over as much as 20 percent of the body. And yet life still flickered in this boy, if just barely. He was unconscious, of course, in severe shock, pulse rapid and faint, his breathing almost imperceptible, and vital body fluids oozed unchecked from the raw wounds all over his body. If Johnson had reached the hospital in such condition perhaps a quarter of an hour later, doubtless he would have been written off as hopeless. Because he was an early arrival, however, he was treated with the other injured: his deep burns were sprayed with triple dye, he was given intravenous infusions of blood plasma and saline solution, and he was assigned one of the beds in the ward upstairs. Even so, nobody believed he could live through the night.

Only a little later that nightmarish evening, when the accident room had already been swamped and the clearly hopeless cases had to be laid aside with the dead to make room for those who could be saved, another startling discovery of life had a particularly moving effect on the City Hospital staff.

Medical personnel were sorting the mostly lifeless bodies already blanketing the floor of the accident section when an orderly stopped short before one. Bending over the still figure, he called out, "Here's one breathing." And as other aides picked their way toward him, the orderly gasped, "Oh, Jesus—it's Dr. Bennett!"

Gordon Bennett had interned at Boston City for more than a year out of Harvard Medical School before moving on, just months earlier, to residency at a suburban hospital. The bright, rugged young doctor had been well liked at City. He looked hardly different now than he always had; the dark, handsome square-jawed features were unmarked, and he appeared almost to be resting. But his skin was ashen and his respiration dangerously shallow. He was lifted gingerly from among the battered, inert forms lying around him, put on a stretcher, and rushed up to an operating room.

Word got around the hospital quickly. Three of Bennett's medical-school classmates happened to be caught up in this emergency. One, Crawford Hinman, who had gone to Dartmouth with Bennett and later—one of seven from their class of '37 to go on to Harvard Med—had roomed with him in Cambridge, was at City Hospital just then by chance. Hinman, who after his own internship there had taken a residency in obstetrics at Boston Lying-In Hospital, was taking a busman's holiday this Saturday evening, dropping into City on his night off to visit with former associates and friends. When the emergency occurred, he'd been enlisted to help along with any other doctors available. He came onto the accident floor shortly after Bennett had been removed, and a priest acquainted with them both told him about it. Anxiously, Hinman hurried upstairs.

Already at Bennett's side were Drs. John Byrne and George Clowes, Jr., two other Harvard classmates, now City residents. They were grim. Gordon, after his charred suit had been removed and his body scrubbed clean, showed almost no visible signs of fire damage. Yet he remained in a deep coma, his breathing labored and weakening as he gasped for air. Tracheotomy had been performed, an incision made in his throat and a hollow tube inserted to facilitate respiration, but it did not seem to be helping much. Rales from his chest cavity sounded ominously as though his lungs were filled with fluid.

Watching him, Crawford Hinman could not help pondering what had led Gordon, of all people, to the Cocoanut Grove. He knew his friend had no taste for night life; Gordon neither drank nor smoked, and to Hinman's knowledge might never even have been in a nightclub before. His consuming interests were sports and medicine—more recently in reverse order. He'd been an exceptional athlete at Swampscott High and then at Hebron Academy in Maine before going on to Dartmouth, where, though relatively small at six feet and 187 pounds, he'd excelled in varsity football and hockey. At the same time, though his grades were only fair, he'd earned recognition as a leader in student government. Gordon had exhibited the same determination in pursuit of his medical degree, which he felt he owed in part to his father, a veteran general practitioner who'd scrimped to send him both to college and to med school. For recreation, Gordon would always be strictly an outdoorsman and a sports buff; to him there would never have been a question of choice between stepping out dining and dancing or attending a hockey match at the Boston Garden—not even on a date with Edith.

Edith. Hinman considered Gordon's longtime sweetheart, perhaps the only one he'd ever had. They'd been a twosome since adolescence in Swampscott. While he was away at Dartmouth, she'd gone to Simmons College in Boston, then to the Katherine Gibbs School—preparing herself, it had always seemed, to marry Gordon. Hinman knew Edith Ledbetter as a lovely, outgoing young woman, quite independent in all matters but those concerning Gordon. Everybody considered them "engaged," though there had been no formal announcement and they had set no definite wedding date. They couldn't yet, not on Gordon's present salary of $35 a week; both knew they would have to wait until he finally could establish at least a modest permanent practice—that's the way they were with one another, patient, confident, practical. All the more reason to wonder what had prompted them so uncharacteristically to splurge on an evening at the Cocoanut Grove. Surely Gordon must have been with Edith; Hinman could not imagine him going to a place like that without her. Where was she, then? What had happened to her?

Hinman found a phone and called downstairs to ask if

any Grove casualty had come in identified as Edith Ledbetter. After a nervous wait, suffering the unremitting confusion below, he was told not yet—or not that they were yet able to tell.

More sorely troubled now, Hinman returned to his friend's bedside. The others, in muted, short monotones or strained silence, tested Gordon, ministered to him, listened to him, watched him, but he had not stirred. His life seemed to hang by a fragile thread. There was little more they could do for him. His parents should be notified. And what about his fiancée?

What a terrible waste, someone muttered bitterly.

The three young doctors knew they could not keep their personal vigil indefinitely. There was too much to be done below and elsewhere in the hospital. They left Gordon Bennett dolefully, each fearful it was the last they would ever see of him, yet each praying silently in his own way.

As the dreadful news flashed across the region, civilian volunteers had begun to flock to City Hospital, where, reports said, the greatest number of Grove casualties were being sent. People from all walks of life, from anonymous, concerned, working-class citizens to wealthy contributors and committee members—some still dressed in evening clothes—came in droves to offer what unprofessional relief they could to the beleaguered staff. They did the most menial tasks, from cleaning and running errands to removing soiled linens and discarded clothing, as well as the most appalling, helping to transport the injured or bed-sitting the critical and the horribly disfigured. (Notable among such volunteers, and yet typical, were the Fiedlers, Arthur, celebrated conductor of the Boston Pops—also widely known as a fire buff—and his wife, Ellen, who rushed back to help from Thanksgiving holiday in New Hampshire. Mrs. Fiedler's brother, Dr. George Bottomley, was an attending physician at City Hospital, and his wife, Lydia, of Boston's distinguished Fuller family, was a tireless Red Cross Aide.)

But no amount of assistance seemed enough: City Hospital found itself simply overwhelmed. In little more than an hour from the time the first fire victims were brought in, well over three hundred had been received.* Soon the two wards in Dowling set aside for disaster patients had been filled to overflowing and new arrivals had to be sent wherever beds were available—ultimately, to over thirty-one wards in all eight main buildings. What's more, supplies were running out: painkilling and antibiotic drugs, plasma and whole blood, the chemical dyes, such basic tools as syringes, bandages, even clean bedding. The situation had become impossible.

At 11:30 P.M., City Hospital administrator Dr. James Manary, in desperation, telephoned the police, Civil Defense headquarters, anybody he could think of with some authority, pleading that no more injured be sent to Boston City.

---

* Probably no hospital anywhere, ever, had received more casualties in so short a period of time. It was calculated later that in an hour and a quarter on average one Grove victim had arrived at Boston City Hospital every eleven seconds. No hospital even in London, England, during the worst of the Nazi air raids had had to handle comparable numbers.

chapter 8

Boston and environs abounded not only in topflight universities and colleges but in first-rate medical facilities, enhanced in no small degree by the advanced teaching and research programs of such schools as Harvard, Tufts, and Boston University. Fine hospitals, large and small, were scattered throughout the city proper and in adjoining towns—Peter Bent Brigham, Massachusetts Memorial, Carney, St. Elizabeth's, Beth Israel, Faulkner, to mention a few—as well as another handful at military installations. The biggest and busiest in the most normal of times, open to all regardless of financial means, was Boston City. But perhaps the best known, and almost certainly the most esteemed, was Massachusetts General.

Mass General, as it is known (literally the world over), was the Ivy League, the Ritz, of Boston hospitals. Chartered, like Boston City, as a "general" hospital, it was theoretically available to all; but in practice such was not commonly the case. It was more a hospital for the relatively affluent, and indeed its patients came from all over the United States and abroad to avail themselves of its expertise.

Boston City, with its own excellent but too often overworked, underequipped, and underpaid staff, tended to envy the better-endowed Mass General. If Mass General, established in 1811, New England's oldest voluntary nonprofit hospital (and the third oldest in the United States), was the Brahmin of Boston's sociomedical establishment, City was its struggling masses. Indeed, City had been founded late in the nineteenth century rather as the poor man's alternative to the elitist Mass General, a haven specifically for the wave of immigrants, mostly Irish and Italian, by then inundating Boston.

The differences between these two preeminent hospitals were both superficial and integral. City was a gray jumble of stone and concrete located off lower Massachusetts Avenue in the teeming south side, an area decaying from shabbily genteel into ghetto. Mass General's setting on the more desirable north side was almost parklike by comparison. Situated alongside the Charles River overlooking Cambridge, close by fashionable Beacon Hill and the Back Bay, it was dominated by the colonial Bulfinch building, a landmark of early American architecture by Charles Bulfinch (also designer of the magnificent State House nearby). Mass General had separate buildings, or "pavilions," of semiprivate and private accommodations for its more selective clientele, notably the exclusive Phillips House (where, it was said sardonically at City Hospital, the patients' regimen featured dry martinis each afternoon on the dot of 5:00). The tiled wards at City, with twice the number of beds, were crammed with people who, often as not, were unable to pay for anything. There, while department heads and senior physicians were of superior caliber, the bulk of medical care was provided by young residents and interns and student nurses who, as a result, received an invaluable training. Mass General, which prided itself on being an international referral hospital for the most complex illnesses, was heavily committed to research excellence as well as to a staff composed largely of proven experts in all branches of medicine. (As the original teaching hospital of the Harvard Medical School, Mass General also had traditionally maintained a distinct preference for searching out its talent from within the Ivy—and Anglo-Saxon Protestant—caste; practitioners at City, on the other hand, were likely to be drawn from other cultural strata and less "exclusive" medical schools, such as Boston University.)

There were at least two aspects, however, in which City Hospital and Mass General were alike. The first was that each was adjacent to one of Boston's two municipal morgues: The larger, Southern Mortuary, was behind City Hospital, across Albany Street at the corner of Massachusetts Avenue; the Northern Mortuary was virtually on the grounds of Mass General, in front of its receiving entrance off Cambridge Street. The second similarity was that, fortuitously, each hospital happened at the time to be vigorously engaged in burn-treatment research under auspices of the federal government.

The scholarly research team at Mass General in fact felt it had at last developed the much improved method of burn therapy so eagerly sought for a world at war, and was waiting for the opportunity to test its findings in a full-blown crisis situation.

It had been natural enough, in the immediate pressure of the great emergency, for authorities frantically trying to organize the rescue effort to direct the mass of victims to City Hospital—Boston's largest, most accessible, most public. But even before Dr. Manary's plea for relief, Red Cross officials at the fire scene had come to recognize, as the staggering extent of the damage became increasing apparent, that no one facility, whatever its capacity, could possibly absorb such an onslaught.

And so the victims began to arrive at various other hospitals across the city. Uncounted numbers of servicemen and women recovered from the Grove by Army and Navy rescue teams were delivered directly to military installations in the area. Most of the other civilian hospitals, however, received only a few patients. The major share of the "overflow" wound up at Mass General, which by 12:45 Sunday morning had received 114 of the casualties.

As extraordinary a rush as that was of unexpected emergency cases to any hospital, the numbers there were not so great as to produce quite the feeling of helplessness that had threatened to overwhelm City Hospital. Mass General had a little more time to gird itself, in any event. And further, its emergency staff was freshly primed for a major disaster, having only several days earlier conducted a thorough "dress rehearsal" of procedures for receiving mass casualties following an enemy air attack. So the sudden crisis there was less disruptive and could be dealt with in somewhat more orderly fashion. There was not even a full mobilization of available personnel.

Student nurses in residence whose next tours were to begin at 7:00 A.M. Sunday, for instance, were neither summoned Saturday night nor informed officially of the crisis until they reported for duty the following morning. (In fact, of course, it would have been virtually impossible for them *not* to be aware of what was going on. One of the young nurses returning from a date awakened the others to report that after dinner she and her Navy man had tried to get into the Cocoanut Grove but couldn't because it was too crowded. On the way home, they'd heard about the terrible explosion that occurred practically just after they'd left.) So while all soon knew, more or less, about the calamity, it was not until the early-morning shift came down to find the floor of the hospital's "great corridor"* filled almost wall to wall with neatly arranged rows of corpses that they finally understood how bad it really had been.

Disparity of total volume aside, the impact was hardly less dramatic at Mass General than it was at City Hospital. Their air-raid drill had not prepared them for this: from all prior recorded experience with disaster, either military or civilian, in a bombed city or even on a battlefront, casualties

---

* Also known as the "red brick corridor," it was actually a vast central room, about 125 feet square, that linked several wings of Mass General.

were anticipated in a ratio of from three to five injured for every death; here, the numbers were reversed. Of the 114 Grove victims delivered to Mass General within two hours of the fire, only thirty-nine were still alive to be treated; seventy-five others died before anything could be done for them, and most of these were already dead by the time they reached the hospital. Again (as at City and the other hospitals), it was troubling how few of these fatalities appeared to have been caused by burn damage.

While the dead were laid out in the great corridor, to be identified where possible and then removed to the Northern Mortuary, the living were speeded upstairs to the sixth floor of the main White Building, where all resident patients had been evacuated to other wards and rooms.

Among the thirty-nine survivors were three married couples, all from Boston suburbs, relieved despite their respective injuries to find themselves still together. (They were, as it happened, the only ones so blessed. Seventeen other couples were pronounced dead at Mass General, as well as another engaged pair who, it would be learned, were to have been married only two weeks hence.)

Most affecting perhaps were the anguished survivors who begged for some information about relatives or friends from whom they'd become separated. Two of these were women from Keene, New Hampshire; one was in her late teens, the other middle-aged. They were from two families who had come down from Keene in a party of six for a weekend on the town—the big football game Saturday afternoon, dinner and a show at the Cocoanut Grove that evening. It was a dual occasion: the older woman's son was celebrating his nineteenth birthday, and in a few days he was to leave for military service; the girl, who was with her parents, was his sweetheart. They'd seen the first floor show and the young people wanted to stay for the second. But when it was delayed the two sets of parents persuaded them to leave; the boy's father and mother and he and his girl went out to get their car, parked in the Grove lot, while her parents waited at the checkroom for all their coats. The four had just gotten out when the fire erupted. The boy and his father battled their way back into the blazing foyer to find the other couple; the two women tried to follow but were hurled back and overrun in the stampede at the main entrance. They hadn't seen any of the others again....

Attendants at Mass General would find out during the night that all four missing—the boy and his father, the girl's father and mother—were among the dead downstairs. They wouldn't inform either of the women right away. The two needed rest, all the strength they could recover to withstand such crushing news.

Others of the injured at Mass General also had been separated from spouses. One was Goldie Yarchin, wife of prominent Boston insurance executive and civic leader Abe Yarchin, who had been in the entourage feting Buck Jones all evening. The last Goldie remembered before coming to briefly in the hospital was the burst of fire, the mass panic, the lights going out, Abe throwing himself on top of her on the steps of the Terrace—Where was Abe? Did *he* get out? She blacked out again before they could tell her they didn't know. (Her husband was, in fact, among the injured taken to Boston City Hospital. Neither would know for days that the other had survived.)

Another was journalist Martin Sheridan, Jones's publicity agent for the Boston visit, who also had been felled in the melee while trying to protect his wife. Dimly aware of having been dragged out by someone, in a fog of confusion and pain, he'd been brought in mumbling feverishly about "Connie" and placed under sedation. They could only have told him then that Constance Sheridan was not yet at Mass General, alive or dead. (She was, it would be learned, dead and at the Southern Mortuary.)

Buck Jones himself was in another bed on the sixth floor of Mass General's White Building. The cowboy star had been found unconscious in a heap of bodies near the Terrace. He remained comatose, his condition critical. The ruggedly handsome features so familiar to more than a generation of moviegoers were swollen like a monstrous pink melon encrusted with patches of black. In addition to burns of the face and neck, Jones had inhaled considerable smoke and fumes, and from his labored breathing it was evident that his lungs were badly seared. (As with each victim, and especially the most doubtful, efforts were made immediately to contact next of kin—in Jones's case, an undertaking made more difficult by the fact that his wife and married daughter not only were in far-off California but, as might be expected of a "celebrity" family, were hardly to be found listed in the phone book. In the end, hospital aides had to contact local newspapers for help.)

The doctor in charge of Mass General's special burn project had hurried to the hospital from his home in Cambridge within a quarter hour of having been notified of the emergency.

Dr. Oliver Cope was more than head of the project: he was both its progenitor and its champion. At thirty-nine, not only learned but commandingly eloquent—and tall, fair and handsome to boot—Cope was regarded as rather a *Wunderkind* by many at Mass General, where he'd been for seventeen years, ever since he was awarded a research scholarship while still in his second year at Harvard Medical School. Now an esteemed endocrinologist (an internist specializing in the glands) as well as surgeon, for some years Cope also had been intrigued by the shortcomings of burn therapy. He had experimented off and on with finding a faster and more effective means both to achieve surface healing and to seal in the vital body fluids that tended to exude from deep burn damage. A paper he presented before the Society of Clinical Surgery in November 1941, citing growing evidence for the importance of containing secretions from the adrenal glands of cortisone and epinephrine ("adrenaline")—the human body's essential response to trauma and shock—had gained wide attention.

When the United States, shortly thereafter, found itself at war, the National Research Council invited Cope, representing Mass General, to an emergency meeting in Washington, where grave questions of wartime medical treatments were to be examined. And there the youthful doctor—youngest in attendance—had proposed a new, simplified, and rather revolutionary method of burn therapy which so interested the council that in January 1942 it had awarded Mass General a contract worth $100,000 to pursue the research initiated by Cope.

The standard deep-burn treatment in 1942 basically consisted, as it had for some fifteen years, of coating burns with a solution of tannic acid to create a leathery eschar, or scab, over the wounds to act as a protective shield against both loss of internal fluids and invasion of harmful bacteria. When first introduced in 1926 at Detroit's Henry Ford Hospital, the procedure was hailed as a signal advance in an area of medical technology that had remained woefully primitive. (Subsequently there had been a refinement to the "tanning" process devised by eminent burn researcher Dr. Robert Aldrich of Johns Hopkins, who had since become a member of the staff of Boston City Hospital. This was the application of so-called triple dyes to burns—the antiseptic chemicals brilliant green, acriviolet and acriflavine.) Most physicians still believed in the therapy as the best modern science could do.

But by 1942 a growing number, Cope among them, had come to question the method's true efficiency and its practicality as well. Under optimum conditions, it was a cumbersome, time-consuming procedure that required complex materials not often readily available in emergency circumstances and a number of technicians knowledgeable enough to administer it—both factors suddenly of particular concern in a time of fast-moving modern warfare. Then there was the question of the steps preliminary to the tanning, which were invariably excruciating to a patient: first, *débridement*, the painstaking and agonizing manual removal of all bits of foreign and destroyed matter from the surface of the wound; then harsh scrubbing and cleansing of raw tissue to prepare for the application of the chemical dyes. Cope and others questioned whether all this handling truly safeguarded a deep burn wound against infection, or whether it might not *encourage* invasion of bacteria instead; and further, they wondered whether such extensive surface treatment—at the expense of neglecting the body's intrinsic need for sustenance and the replacement of vital fluids—might not actually promote shock rather than helping to avoid it. (Shock was the most lethal danger in such trauma cases, and was commonly the result of irreplaceable loss of body sera; but only in recent years had the medical profession begun serious research into this critical area.)

In essence, what troubled Cope, for one, was the *artificiality* of the whole laborious "tanning" therapy. He felt that the human body possessed formidable regenerative powers that all too often were not taken into account by medical science. If particular care was taken to promote a *natural* healing of burns, might not treatment and recuperation be accomplished more simply, quickly, and safely? The bleb, or blister, formed over a low-degree burn seemed to Cope an applicable case in point. As long as it remained unbroken, the bleb was the best body-produced guard against both outside infection and loss of fluids from within: in a brief time, oozing sera were reabsorbed into the tissues, dead cells formed a crust over the damaged area, and, when the blister fell off, the wound was healed.

It had been in 1932 that Cope—then a young general surgeon only a few years out of Harvard Medical School—first found occasion to implement this concept, which until then he'd only been able to consider in his few idle moments as a staff physician at Mass General. One February morning he got a frantic call at the hospital from his sister in suburban Belmont: her three-year-old daughter had overturned a steaming pot of coffee and scalded a hand and arm. Cope told his sister to immerse the child's burns in cold water until he got there. On his way, he appropriated from the emergency room a large jar of petroleum jelly and a package of sterile gauze.

He found the little girl reasonably calm, her hand and wrist dangling in cold running water at the kitchen sink, her mother trying to read quietly to her. Blebs already had formed on the reddened skin; the burns were serious, but no worse than second degree. Cope gently dried the wounds, then daubed them with the petroleum jelly, being careful not to break the blebs. Finally, he covered the whole area loosely with the gauze and secured this dressing with a second layer. With a scarf of his sister's, he devised a sling and positioned the girl's bandaged arm high up against her chest. Before leaving, he admonished his niece—with a kiss—and her mother not to disturb or remove the dressing for any reason until he came to examine it again.

When Cope later told his superior at the hospital how he had treated the injury, the older physician was surprised and dismayed. It sounded to him as though Cope had experimented unnecessarily with the child; he should have followed normal procedure and tanned such a burn. But Cope insisted his method would work. "Wait and see," he said.

The girl's hand did heal, thoroughly and quickly. Within ten days, fresh epithelium (new skin) covered the surface area of the burn. The skin remained red, gradually diminishing, for several weeks, but in the end it would be almost impossible to detect where or what the damage had been.

Between that time and November 1941, Cope, while primarily occupied with his endocrinology practice (and gaining increased recognition for his work, particularly in the field of thyroid diseases), had found random opportunities to treat other burn patients by the same simple method as used on his niece. Each case had proved equally successful... and yet, as sure as Cope was that he was on to something, there had remained a certain reservation: none of these cases had yet involved worse than surface or second-degree burns; still to stand the test were those of the disaster type, where direct flame damage destroyed all layers of skin to expose raw tissue, internal organs, bone. Then, most critical would be the effectiveness of his treatment not only in defending against infective bacteria but in containing precious body fluids, and providing for the rapid replenishment of sera already lost.

Thus, when the National Research Council met in December 1941, Cope presented his theory, and only weeks later he and Mass General received their experimental grant.

In the ensuing ten months, Cope and a select group of chemists and bacteriologists had treated seventy-three special burn cases, and in each case the results had been positive: without exception all their burns had healed completely, not one patient's bacterial count had gone above zero, and none had gone into shock. By November 1942, the hospital's directors were sufficiently impressed with the soundness of Cope's treatment to decree it the new burn policy for both of Mass General's surgical services.*

Now, this late November night, Cope had been summoned from home to assume management of a far greater burn crisis than he or any of his associates could ever have imagined. On duty, he was gratified to note, was the East Surgical Service with which he'd worked most closely in formulating the new burn therapy. He had no less confidence in the West Service, which had recently been instructed in the procedure and stood ready as backup, but the kinship of previous shared experience was reassuring in the emergency circumstances. The East Service was already in gear when Cope arrived.

Of the thirty-nine living admitted to Mass General, ten were found to have sustained significant burn injuries. Each

---

* Mass General had two surgical services, dubbed East and West; Boston City Hospital, in contrast, maintained five.

was first injected with morphine sulfate. Sopping wet and filthy clothing was removed, and undamaged skin grimy with soot was cleaned, but not the blackened burn areas. The untouched wounds were covered loosely with fine-mesh gauze impregnated with boric petrolatum ointment, wrapped in turn with cotton and more gauze and secured with elastic bandaging. Pressure dressings were applied only to areas of the head and extremities. To counter shock and lowered blood pressure, saline solution, blood plasma, and liquid protein were administered intravenously, and later sulfadiazine and antitetanic serum for infection. Eyes were examined by a resident specialist, who found corneal lesions caused by heat and smoke in seven of the victims and applied sulfathiazole ointment and atropine drops.

The other twenty-nine survivors had relatively slight burn injuries, some none at all, yet a number of these also were in grave condition—undoubtedly from severe pulmonary damage and anoxia (restricted oxygen supply to body tissues). Most were quiet and pliant, stuporous; but several displayed various stages of hysteria, reacting irrationally and even violently to all attempts to minister to them, and requiring physical restraint by orderlies. (Caution was exercised in administering morphine to these lest the narcotic fatally depress already disordered respiratory systems.) There was frequent vomiting by the smoke victims.

By 3:00 A.M. Sunday morning, all the patients were bedded down on the White Building's sixth floor—twenty-one males grouped on one side of the building, eighteen females on the other. Thirty-two were judged to be in good or stable condition. Seven were listed as critical, among them Buck Jones.

Through the long, black night, across the Mass General compound, in a private room in Phillips House, isolated from the turmoil swirling all about him, lay a heart patient who would be profoundly affected by these ghastly events. He was Barnet

Welansky, successful attorney and proprietor of the Cocoanut Grove. He did not yet know of the devastation to his club at the cost of so many lives. But when he learned of it, in a matter of hours, it came as a crushing blow from which he never fully recovered.

chapter 9

How many casualties were there?

For many hours, no one had been able to say for sure—or had dared to guess.

The first bulletins, flashed across the nation and worldwide by the wire services and radio, had estimated "hundreds injured" and a probable death toll of "scores." It was the Boston newspapers, with virtually their entire staffs mobilized to cover every aspect of the unprecedented local disaster, that began turning scattershot guesswork into more or less reliable computation.

Reporters and photographers from all the metropolitan dailies* as well as from the suburban and regional papers, radio correspondents with their live microphones, and newsreel cameramen hefting bulky equipment swarmed everywhere about the fire scene. Every few minutes the print reporters would scramble to neighborhood bars or wherever else a telephone could be located to call in to rewrite desks their breathless, horrified descriptions of the carnage, as well as comments snatched on the run from rescuers, dumbfounded

---

* Seven daily newspapers were then being published in Boston: the *Record, American, Herald, Traveler, Globe, Post,* and the international *Christian Science Monitor*.

survivors, awed witnesses, and harried public officials; messengers on motorcycles raced like Pony Express riders back and forth between the Grove and their photo labs with Speed Graphic plates capturing the devastation and the misery.*

Others descended upon the hospitals and mortuaries. They dogged doctors and nurses, technicians and non-medical personnel, even cornered the incoming injured, scribbling names and addresses and ages and whatever intimate accounts they could scrounge. They haunted the rooms full of dead and the mortuaries; they counted; they accosted grief-stricken survivors.

It was not an easy assignment; exciting, but not exhilarating. Most of the newsmen were as shocked as anyone else by the unthinking horror—many had covered fires with fatalities before, but never anything close to this. If they seemed intrusive—and some were—it was the unfortunate nature of their jobs. They could not get information without intruding on others' pain and despair. (The worst responsibility perhaps fell to those sent out in the night to the homes of victims all over the Boston area to ask shattered families for snapshots or photos of their lost loved ones. It would take many a reporter, even the most senior, a long time to rid himself of the feeling of debasement that came with such tasks.)

They got their information and their pictures, however distasteful, and inundated their newsrooms. And so the papers were able to compile, long before anyone else, an accurate profile of how immense the tragedy was. Rushing out Sunday editions early, each paper competitively replated its front pages with successive updates of total casualties, especially the fatalities: two hundred... three hundred... revising upward

---

* A veteran photoengraver at the Boston *Globe* fainted upon seeing the first raw film of the dead and horribly injured. Few such blood-curdling photos would actually be published, however, as newspapers imposed on themselves a self-censorship against tasteless exploitation. The most shocking pictures, indeed, were filed away under lock and key—and have remained so to this day.

through the night—all the way up to the *Globe*'s great bold headline in its early-morning "final": *450 FEARED DEAD.*

Some, notably city officials, had at first quarreled with such numbers, denouncing the newspapers for irresponsible sensationalism: It *couldn't* be that bad! But by morning they knew better. And by then there were more than a few who dreaded that the *Globe*'s terrible projection might even fall short.

Within two hours of the holocaust, Boston City Hospital had admitted 132 injured victims of the Cocoanut Grove and Mass General thirty-nine. Many of these required intensive care, but almost from the start the facilities of both institutions had been far more sorely taxed by the staggering numbers of dead. Just these two hospitals by after midnight had recorded some three hundred dead—seventy-five verified at Mass General, over two hundred and still counting at City.

Each hospital had begun making up its own files of vital statistics on the victims as soon as they came in, alive or dead. This was more than mere administrative paperwork: it was necessary preparation for the anticipated deluge of inquiries they would have to handle.

And come the deluge did, starting soon after the first radio bulletins. People descended in waves upon the hospitals, chiefly City and Mass General—fathers and mothers, sons and daughters, other relatives, neighbors, co-workers—all frantic with anxiety and bewilderment. Those who could be told that dear ones were injured but alive went away weak with relief (visitors at that critical stage were permitted only to those patients whose prognoses were doubtful). Many more had to be given the heartbreaking news of those who had perished, identification confirmed. Some, grasping at the tiniest wisp of hope, asked to see the remains; but there never was a mix-up, and they would drift off in a stupor of uncomprehending grief.

But at least, while it was hardly a consolation, these people *knew*. More pitiable were those who came and found *no* listing of the ones they sought, alive or dead. They would experience initial spasms of relief, even hope...followed, inevitably, by renewed torments of doubt. Had their relatives or friends got out safely? (Maybe, for some reason, they'd not even been there after all!) Or... were they among the still unidentified dead?

Many of these anguished searchers roamed the city through the night, going from hospital to hospital and haunted mortuary. And the names they sought in vain were added at each repository to a lengthening roster of the "missing."

As difficult to handle, and sometimes even more heartbreaking, were the phone calls—equally urgent, pleading, hundreds and hundreds of them, endlessly through the night. Many were from the worried families of college students and servicemen, and they came from all over the country—to the hospitals and the Red Cross, to City Hall, to newspapers and radio stations, begging information. But only a few were able to learn anything definitive. No hospital, embroiled in its own immediate crisis, could account for what was happening anywhere else. The problem, in the early chaotic hours, was that there was no centralized list being kept of identified victims.

It was not until early Sunday that a unified information system finally was organized by the Boston Committee on Public Safety (the wartime agency authorized to coordinate the city's response to enemy attack). Once lines were established into every area hospital and morgue, a cross-referenced "master file" of Grove casualties took shape: name, condition, and whereabouts of all survivors on green file cards, the known dead on pink cards, the growing list of missing on white— more than seven hundred in all. From then on, most inquiries were referred to the central committee switchboard, where operators could pinpoint any name at a glance.

One operator there logged over a thousand incoming calls in a single shift. And still, long after daybreak Sunday, the phones continued to ring.

* * *

In different ways, disposition of the deceased, especially those not immediately identifiable, posed as great a challenge as tending the injured—perhaps greater, considering the disproportionate numbers.

At first, as victims had poured in, those judged to be beyond help were simply laid aside—on the accident floors, in triage examining rooms, in waiting areas and in corridors—as efforts were toward caring for the living. And before long, as the widening rows of the dead began to encroach upon precious working space, they were loaded onto flat cars and hauled to the nearby mortuaries, many without having been "processed."

But as the initial impact of the catastrophe passed, and teams of medical professionals, schooled volunteers, Red Cross workers, police, and military representatives were organized to help, concerted attention was given to systematizing the registration of fatalities.

The first requisite was identification of each body. This was difficult both because of the extent to which some had been ravaged and because so many were women. Most males still carried wallets or other identification in their pockets; almost all the females had left behind or been separated from handbags and outer coats likely to have contained personal effects. In many cases there were only items of jewelry or articles of clothing to provide helpful clues. But there were some who bore no clues at all.

The unidentifiable were examined carefully and file cards listing all observable bits of information were made out: in addition to jewelry or clothing, the height, estimated weight, color of hair (including where possible the true color if it had been dyed), general physique and any physical abnormalities, surgical scars, and dental work—anything that might help cue relatives or friends without their having to go through the ghastly ordeal of picking through row upon row of disfigured corpses.

But it soon became apparent to hospital workers that even the most meticulous cataloguing would be of only partial use at best: it provided clues, but rarely confirmation; each agitated inquirer had to be interviewed, as thoroughly as could be managed in the distracting circumstances; and, if the information seemed to match, somebody in each case still had to view a body, sometimes many bodies, before positive identification could be made—or not. Each visitor (after proving relationship with the deceased) was accompanied by an authorized staff member into the morgue, and there kept under scrutiny by police officers posted to prevent theft of valuables from the dead.

By early Sunday, a majority of the dead had been identified, receipted by next of kin, and scheduled for collection by private undertakers.

But a number—most of them women—remained uncertified, and at dawn, long queues of bleary-eyed people, randomly bundled against the cold, still waited outside each mortuary in search of "missing" victims. They watched in bleak silence as open trucks stacked with new coffins, and then lines of black hearses, began to draw up alongside.

All day Sunday, almost from first light, crowds gathered to gawk silently at the smoldering ruins of the Cocoanut Grove. Scores of policemen formed a perimeter around the disaster area beyond which no unauthorized persons were permitted. Piedmont and Shawmut streets, and the sidewalks off Broadway, remained littered with rubble—broken furniture, piles of shattered glass and building materials, remnants of torn and burned clothing—and structures on all sides of the club were blackened and scarred.

An official inspection began early Sunday afternoon with the arrival of limousines bearing clutches of somber, important-looking men—most middle-aged civilians in dark, heavy overcoats and grey felt hats, but also others in uniform, some set off with conspicuous insignia and braid. They included the governor, the mayor, state and municipal commissioners, agents of the Massachusetts Attorney General's office, the county district attorney and the Federal Department of Justice, Boston fire and police commanders, and ranking representatives of the area's military commands. They poked at the wreckage, ventured gingerly inside in small groups, and sent their adjutants off on quiet missions. They clustered together outside in the filthy gutters conferring, and wondering what it would take to get to the bottom of this awful tragedy.

What would they find? Might heads roll as a result—maybe even their own? All seemed to sense that the investigation was not going to be easy, or pleasant.

During that Sunday afternoon, one nonofficial person was allowed to pass through the police lines to return to the Cocoanut Grove: Daniel Weiss, nephew of proprietor Barney Welansky and the club's undercover "spotter," who had been tending a cash drawer inside the Melody Lounge Bar and who may have been the last individual to escape that room under his own power. The young doctor-to-be (he had already applied, and in four months' time would be accepted, for an internship at Boston City Hospital), in unabashed loyalty to his supportive uncle and to the responsibilities entrusted him, had come back as soon after the fire as he could to inventory any tills left deserted the night before—including, first of all, his own. . . .

When Weiss got out alone via the backstairs from the basement, having failed to persuade the terrified patrons in the kitchen to follow him, he'd almost fainted in the street from the sudden release of sustained tension. He'd slumped against a building wall trying to compose himself, almost oblivious to the firefighters rushing around him. Then he remembered the ones below and started back toward the smashed door he'd just come out, wondering if any had tried to follow after all. But a fireman shoved him aside, and Weiss could see it was madness to try to re-enter the club. All he could think to do was to cry out that people were still alive downstairs . . . and in a few moments he wandered off along Shawmut Street, away from all the tumult.

Weiss didn't know quite what to do or where to go. But he thought of his parents and how worried they would be if they heard about this, and it came to him that the first thing they might do was try to reach Uncle Jimmy, his mother's younger brother. So, hardly aware of having reasoned this out, he found himself walking the few blocks to the Rio Casino over on Warrenton Street.

Jimmy was not there; but some of the Grove's help and entertainers were, and Weiss, though he was not particularly close to any of them, nonetheless felt reassured being among them. He telephoned home, to the immense relief of his family, who had been frantic. And while they hurried from Roxbury to pick him up, Weiss at last unwound over a cup of coffee.

By the time his parents got there, word had come in that Jimmy Welansky was safe, unhurt. Now their one familial concern was for Barney: bedridden and helpless with a diseased heart, could he withstand learning of the calamity?

Daniel's father, who was himself a doctor, in hugging his son noticed blood from a cut on his neck. Dr. Weiss insisted that it was too close to an artery to be taken lightly, and that, in any event, Daniel should be examined thoroughly for any other injuries that might have been numbed by the shock. They took him to City Hospital.

It was after midnight by then, but the admitting floor was still more disorderly than Weiss had ever imagined it could get, though, of course, he had no idea how chaotic it had been only a short while before. He found a friend of his, another graduating medical student, staffing the crowded emergency desk. The drawn young man did a double-take on recognizing Weiss. "Oh Lord, you too?" he groaned.

"Just a scratch," Weiss said apologetically. "Nothing compared to—" He looked about in dismay. "Is it very bad?"

"The worst. Unbelievable," said his friend with a grimace. "Count your blessings. When I saw you, I—" He paused to write Weiss's name in the admitting book. "A little while ago

111

they brought in one of the doctors, used to intern here. Young guy, big, ex-football star at Dartmouth, I think. Gordon Bennett. Know him?"

"The name is familiar."

"You should have seen him. They thought he was dead. And he was all but. He'll never make it, I don't think."

"He was at the Grove?"

"Yeah. I guess that's why, when I saw you... you know?"

Weiss was hurriedly examined head to toe, and they found another laceration he didn't know he had, a gash on a leg that had already stopped bleeding. Both wounds were cleaned, judged to be superficial, and dressed lightly. Weiss was given an antitetanus shot and his parents were told to take him home and make him rest.

The enormity of it all did not really sink in until Sunday, as the full extent of the tragedy became known and the telephone began to ring with anxious inquiries from relatives and friends wanting to know if Daniel was all right. In a perverse delayed reaction, his parents, instead of being limp with relief, seemed to grow more agitated as the day wore on: his mother, unable to sit still, moved distractedly from room to room wringing her hands and moaning—"All those poor people. Poor Barney"—while his father now began to chastise him over and over for having caused them such heartache—"How many times did I tell you? A nightclub is no place to work—not for a doctor!"

Barney Welansky was prominent in Daniel's own confused thoughts—how much his uncle had tried to do for him, *had* done, even if not in all the ways Barney had wanted. Daniel knew what some people thought and said of Barney: that he was a cold fish, a martinet whose only love was for the buck, an ethically shady opportunist who could cultivate favor in high places, like City Hall, and as easily among big-time racketeers, to achieve his mercenary ends. But this was not the uncle Daniel knew.

Daniel had worked his own way through college; his father's income as a GP had been meager during the Depression (when on many a house call he was more likely to leave a dollar or two than to receive payment), certainly too modest to permit such luxuries as advanced educations for a son and two daughters. When Daniel considered trying for medical school, Barney, whose marriage had produced no children, offered to subsidize him. But Daniel refused his uncle's generosity, insisting on earning his own way as he always had, and Barney had settled for buying him basic items of equipment—a microscope and such—which he would present on birthdays and any other occasions he could manufacture. At the end of Daniel's first year at med school, his resources were falling short and, discouraged, he announced he was thinking of quitting and asked if Barney could find a spot for him in his nightclub business. Barney said he would *give* him what money he needed. Daniel again said no, he didn't want it that way, and no loans either that he didn't know when he'd be able to pay back. Instead, they struck a bargain: if Daniel would determine to stick it out, finish school, become a doctor, Barney would agree to hire him weekends and holidays at the Grove. And so, thanks to Barney Welansky's own determination and wisdom, his nephew had now almost completed med school.

And now, too, everything Barney Welansky was and had might have been wiped out. Daniel didn't know much about the Grove's finances, but he doubted insurance would come anywhere near covering such monstrous losses of life and property. Could Barney *ever* recover from such setbacks— could anyone? But even if it was hopeless, he was going to need every bit of help he could get. If Daniel owed his uncle nothing else, he felt, he owned owed him whatever support he could give.

Thus did Daniel Weiss return to the Grove hardly more than twelve hours after the fire to see how much, if any, of Barney's cash he might retrieve from the ashes. A drop in the bucket maybe, but if there was *anything*...

The place looked gutted, everything black and broken and gaping mournfully. His chest tightened as, accompanied by a fire marshal with a torch, Weiss cautiously descended the cluttered stairs to what had been the Melody Lounge—trying not to imagine the tortured bodies that hours before had been flung across each step he took. Below was eerie. The dank air was redolent with the odors of smoke and seared flesh. Everything was waterlogged. Upended furniture was scattered about. Weiss glanced at the far interior corner where the fire seemed to have broken out: only the top of the fake palm tree was burned away; the trunk, though singed, was intact. Indeed, as he looked about the room it appeared that only the upper parts had suffered any flame damage at all; the walls from about chest height down, the floor, even most of the furnishings were hardly scorched.

Ducking under the bar—the opening clear now of the human obstructions that had trapped him the night before—he located his cash drawer. It was grimy on the outside, but inside the contents were untouched. Weiss scooped the bills and change into the canvas money sack he'd brought, not pausing to count it. Then he went around the bar to the other till. It, too, was undamaged, undisturbed. Eerie. He hurried out through the rear service door into the kitchen.

Except for a residual thin haze of smoke, the kitchen showed little evidence of fire. Perhaps the flames had never reached here or had been unable to sustain themselves on the metal and concrete. Had the knot of fearful people huddled down here found a way out or been saved? What of Katherine Swett, that doughty Irish lady? Weiss turned to ask the silent marshal... then decided not to; he wasn't sure he wanted to hear the answer. The drawer that cashier Swett had guarded so stubbornly—perhaps with her life—was also untouched. Weiss emptied it into his sack and went quickly up the stairs to the main club.

In the show room, the destruction was more assaulting to the senses than below, because it had been so much grander to begin with: there was so much more space, so many levels, such heaps of skewed furnishings. The walls and plate glass windows on both sides of the room had been smashed through to the street, and a yawning hole chopped out of the ceiling let in the jagged daylight. But even here, as Weiss peered about incredulously he could see that most of the fire damage was confined to the upper reaches. It was almost, he thought, as if a work crew might actually make this place habitable again. He shook the vision from his head. Insane. The Cocoanut Grove was gone forever.

There were three more registers on the main floor, one at either end of the Caricature Bar, the last in the New Lounge. (That room looked to be hardest hit of all.) Weiss found each as it had been left. He didn't know how much he had collected, but it must have been several thousand dollars at least. Count your blessings, his friend at the hospital had said. For whatever good it might do, he'd salvaged *something* for Uncle Barney.

He returned to the club room for a last, aching look. His gaze was caught by a flutter of white up on the wreckage of the bandstand. Curious, Weiss picked his way over. What he'd seen were scattered pages of sheet music riffled by wafts of air from the open roof above. Hardly any were so much as singed. He leafed through them. On top was "The Star-Spangled Banner"—since the war began they'd always led off with Billy Payne singing the anthem. In disorder underneath he found "I Said No," "Maria Elena," "Babalu," "We'll Meet Again"... then "I Burned a Match (from End to End)."

Eyes smarting, Weiss turned away abruptly and left.

chapter 10

Katherine Swett, the kitchen cashier, had died at her post. As had the headwaiter Frank Balzerini. Most of the other Grove employees, who, unlike a majority of patrons, knew where the unmarked exits were, had gotten out without serious injury.

Herb Lewis, accompanist to singer Maxine Coleman in the New Lounge, escaped; she did not. Grace Vaughn McDermott, a fill-in singer from New York—the attractive young woman who'd rushed from the main club into the New Lounge to warn of the fire sweeping that way—was overcome there and died.

The alternate performer in the New Lounge, young Dorothy Myles, had been on her break in the powder room off the corridor leading to the main club when the firestorm erupted. Rushing out, she'd been overrun in the stampede and buried by falling bodies. Later, firemen removing the corpses noticed a movement of her arm and extricated her. She was horribly burned about the head and arms and on the critical list at Boston City Hospital.

Of the show-room performers—the last to become aware of the fire but fortunately the farthest removed—all were safe but one: a male dancer, Guy Howard, had tried to get out through the main club and did not make it.

Goody Goodelle, the Melody Lounge entertainer, for a time afterward was listed as missing and presumed dead. In fact, she was relatively unhurt but suffering severe emotional trauma. Goody had led the petrified young female cashier from the Melody Lounge out through the kitchen and up the stairs to the show room. The fire had just begun to rage there, and the two women and some waiters had escaped by smashing out a window onto Shawmut Street. Goody then wandered the streets in a daze for several hours before regaining her senses enough to find her way home. Meanwhile, however, fearful relatives had begun searching the hospitals and morgues. At the Southern Mortuary, in the Sunday predawn, one of her cousins came upon the charred, unrecognizable corpse of a young woman wearing shreds of a brilliant red evening gown such as the one Goody had worn that night, and she collapsed in horror and grief. But by then Goody was home in bed.

One member of the orchestra was dead: the violinist and leader Bernie Fazioli. Caught in the crush at the locked exit behind the stage, Fazioli fell victim to the smoke and fumes before the door could be broken open. He was taken to City Hospital, barely alive, but it was too late.

Two others who had been alongside him, Jack Lesberg and Al Willet, also succumbed there and were rushed to City Hospital. Each had suffered serious pulmonary damage, but they were still alive. The drummer, Al Maglitta, managed to stay conscious and got out when the door was finally opened. Maglitta was treated for superficial injuries at Massachusetts Memorial Hospital and sent home.

Another orchestra member, saxophonist Romeo Ferrara, was also treated at Mass Memorial and released. But he, too, was mistakenly reported as a probable fatality for several hours. When the Committee on Public Safety could not reach Ferrara at home to verify his escape, his name was placed among the "missing" in the master file of casualties. Ferrara himself corrected the error when he saw his name and picture in the papers the next day. As it turned out, his phone had been disconnected just that Saturday because he and his family had moved to a new apartment in Somerville.

Pianist Moe Solomon, Jack Lesberg's cousin, had very nearly died. Pinned against his piano by the mob pouring over the bandstand, Solomon had started to grow faint from the sickening fumes and choking smoke. His will to struggle began to seep from him... and then, distantly, he heard a woman's desperate wail: "My babies, my babies!" With a jolt, Solomon was reminded of his own youngster, and his wife, who needed him... and he found somewhere within himself a renewed resolve, a flickering reserve of strength, and he made himself *move*, stumbling, crawling blindly, until, somehow, he came to an opening to the street, and sprawled out onto a frigid sidewalk, gulping air.*

Trumpeter Henry Minasian was the one who had found refuge in the standup refrigeration locker in the club's basement. Minasian had been substituting just one week for a regular trumpeter, veteran Hy Cohen, and Saturday night was to have been his last performance with the Alpert orchestra. Cohen, expecting to be drafted, had been absent that week to take the Army physical and get his induction date and was due back at the Grove Monday, the 30th.

In the tumultuous hours after the fire, it came to seem that just about everybody in Boston knew or knew of somebody who was at the Grove that night—or who was supposed to be but, by some quirk of circumstance, wasn't; somebody who had cheated death, or someone who hadn't....

---

* In fact, there *was* a baby in the Grove when the fire broke out. One of the showgirls—separated from her husband and without family nearby—had brought her infant daughter to the club and left her bundled in a storage closet off the show room while she went about her night's work. At first, nobody paid any attention when the hysterical young woman started screaming about saving her baby. But finally several patrons did heed her cries, found the child and managed to pass her safely through a smashed window before helping the mother out as well.

And with the tales of horror there also emerged, as if to relieve the crushing despair, redeeming accounts of individual sacrifice, even of heroism, to affirm the fact that even in the worst of times people would still reach out to help one another.

There were the hundreds of outsiders, volunteers who'd flocked to the scene—many of them servicemen barely out of their teens and far from home perhaps for the first time—and who had ignored personal security to help total strangers. Already being talked of as a folk hero was the baby-faced Coast Guardsman who had returned time and again into the flames to pull trapped patrons from the flames in the New Lounge, only to collapse at last, so terribly burned that surely he must have died. (Miraculously, the sailor, Clifford Johnson, was still alive at Boston City Hospital.)*

And the indefatigable clergymen, mostly priests from the heavily Catholic South Boston district, who'd come out to minister through the surreal night to the injured and dying, often at great risk to their own safety—if also, at times, to the consternation of beleaguered firefighters, most of *them* Irish, who, stumbling upon a crouched, murmuring cleric had no recourse but to rebuke him as respectfully as urgency allowed: "Jesus, father! Excuse me—but you're in the goddam way!" (The priests, with no chance to determine the victims' religious faith, conditionally baptized and absolved any they

---

* One who narrowly escaped death was the sailor boyfriend of Grove dance captain Jackie Maver, whom she had been supposed to meet after the last show. Al "Whitey" Drolette, waiting for her at the Hotel Avery bar, heard the early radio reports of the fire and raced to the club to find her. He tried several times to enter the blazing building before finally succumbing to the smoke and flames. Believed dead, he was deposited with other corpses on the floor of a garage across the street. When he was finally discovered to be still alive, Drolette was removed to the naval hospital at Chelsea, north of the city. It would be thirty-six hours before Jackie, frantic at having had no word for him, learned of his whereabouts from the official casualty lists. A melancholy postscript: Drolette spent many weeks recuperating at Chelsea, then returned to his Navy unit and shipped out. He and Jackie never got together again.

could reach. Many non-Christians later would recall, with a certain bemused reverence, having roused to find themselves being anointed and croaking: "I'm Jewish—but please, father, go right ahead!")

There were taxi operators and truck drivers, both voluntary and conscripted, who shuttled back and forth between the Grove and hospitals all over the city until the ambulance services could be organized. And the hotels nearest the club, the Statler and the Bradford, that sent linens and sheets and other provisions to the fire scene and even opened their ballrooms and unoccupied guest rooms to exhausted survivors and rescuers.

There were the blood donors who responded so unhesitatingly when word went out, only hours after the disaster, that the hosptials already were running out of vital plasma as well as whole blood. On Sunday alone, 1,200 citizens offered blood, and within days almost 3,800 units were collected by the local Red Cross.* (One group of young "conscientious objectors"—men whose religious beliefs or ethical scruples forbade their serving in the military—in the darkest hours of Sunday morning went out onto Charles Street, the busy thoroughfare in front of Mass General, to flag down passing automobiles and plead for blood donations. They succeeded in bringing several dozen surprised motorists into the blood bank. Unfortunately, many of these contributions were unusable because of the high alcohol content.)

There were the many hundreds of private or retired nurses and nurses' aides who responded to the urgent call of the Red Cross, not only to augment overwhelmed hospital staffs but to provide home care to the survivors and bereaved. They would work for days and weeks and in some cases for months

---

* The Red Cross also organized airlifts of blood from Washington, D.C., New York, and other New England cities. In the first 24 hours, more blood plasma was administered to Cocoanut Grove victims than had been used in Hawaii immediately following the attack on Pearl Harbor.

afterward with minimum or no compensation.

And, of course, there were the duty personnel at the hospitals, particularly the two main ones—between them, Boston City and Mass General received about 80 percent of the casualties—the doctors and interns and nurses and student nurses who worked around the clock for two or three days without relief. (The effect on one of these was typical of many. It was not until Monday that young nurse Anne Montgomery, who had been on duty at City Hospital since 7:00 A.M. Saturday, finally was ordered to go home to snatch a few hours' rest. Still wearing her soiled white uniform, she boarded an elevated train, lost in a daze of physical and emotional exhaustion. Her eyes would not stay open. But partway home she jerked awake to the realization that other passengers were staring at her curiously. She withdrew into a corner of the seat, self-conscious about her unkempt appearance. And then the odor assailed her nostrils, an odor coming from *her*, clinging to her uniform, her hair—the putrid smell of burned flesh. She got up and moved to the end of the car and sagged against the door and wept. Could any of these people have *any* idea what she'd seen in the last two days? At home, Anne crawled into bed, unable to talk about the ordeal with her family. She cried silently but uncontrollably into her pillow before at last dozing off. And then she dreamed, of blackened flesh writhing in a roiling sea of purple dye, purple everywhere smeared over globs of black—and she woke gagging. It was a nightmare that would torment her for a long time to come.)

And then there was the remarkable coincidence of that secondhand auto, its seat cushions smoldering, that had brought out the first firefighters, as yet unaware of the greater drama already underway so close by. If they had not been in the vicinity, had not happened upon the inferno as they passed, how much worse—without those few minutes' grace—might the final tragedy have been?

Before the long night was over, the battered "mystery" car had been removed, presumably by its owner. But the license number had been noted, and later the vehicle was traced to a family in Brookline. The driver had been a twenty-year-old man, a defense worker who, with his eighteen-year-old brother, a college student, had driven into town Saturday evening for a round of holiday bar-hopping. Finding all the downtown garages filled, they'd parked on Stuart Street just in from Broadway, in front of the Animal Rescue League Building, locked the car (they'd thought), and gone off. By the time they returned at about half past midnight, police barricades cordoned off the area for blocks around the Cocoanut Grove. Thousands of spectators crushed against the barriers, and loudspeakers exhorted them to disperse so as not to impede the rescue work.

The two nervous young men finally persuaded a police officer to let them in to retrieve the car and get it out of the way. They were surprised and upset to find the windows smashed and the front seats torn out. Assuming it to be the work of vandals, they drove home with the older brother using the spare tire for a seat.

It was not until two days later, when fire officials checked the ownership of the vehicle that had prompted the first alarm Saturday night, that the family in Brookline understood what actually had happened. But how had the fire started in the car? Neither brother smoked; and the car *was* locked when they'd left it, they were sure. Had a window not been fully closed, and had some passerby, carelessly or meanly, flipped a cigarette butt through the opening? There was no telling; no one would ever find out. But what was most amazing to them—as they began to see from the newspapers and hear over and over on the radio—was that, not only were they not being blamed for the trouble their car had caused, the incident was being hailed as the providential circumstance that might have saved a hundred or more people from perishing in the Grove!

But with all the uncanny, touching or uplifting stories to emerge from the calamity, inevitably there were also the ugly ones.

In the Grove itself, the locked exits and concealed escape

routes were, if not proof of intentional mean-spiritedness, then unquestionably an example of careless or even callous disregard for human welfare.

And, there was the persistent accusation—sworn to by some who were there, but not authenticated by others—that an unidentified employee of the club, a captain or assistant headwaiter perhaps, had stood adamant in the Foyer, trying to block the way of terrified patrons charging for the revolving doors, proclaiming stoutly: "Nobody leaves here until they pay their bill!"

And there was the beastly, shameful behavior of some of the patrons themselves: embittered survivors told of having been overrun and clawed at, of having seen women and elderly people thrown aside and trampled by others—mostly healthy men!—whose only concern was self-preservation.

And outside, amid the carnage and confusion, jackals and furtive scavengers crept among the dead and injured to strip them of valuables before any sort of security could be organized. (There were dozens of reported cases of stolen wallets, cash, wristwatches, jewelry, furs. Grove saxophonist Al Willet, for one, would discover as he lay returning to life in Boston City Hospital that he'd been so victimized: a finger had almost been severed as his diamond ring was forcibly ripped off; his watch was gone, and also his wallet.) There would be investigations, of course—and pawnshops throughout the Northeast would be alerted to watch for the items reported stolen—but from their experience the police knew that such random crime held scant hope of "solution."

But always, overshadowing the incidental ugliness, was the pervasive pall of death.

One facet of the crisis that increasingly confounded the medical people as the bodies continued to pile up was the seemingly inordinate number of fatalities from *other* than extensive burn damage.

Deaths by suffocation from smoke inhalation or carbon monoxide poisoning were hardly surprising in a fire of such extraordinary volume and evident intensity. What was puzzling was that so many seemed to have died instantly, or after they had reached safety, or en route to the hospital; even more baffling were those who had come into the hospital apparently with only minor injuries or even none at all, and who had, with little or no warning, collapsed and in some instances expired practically on the spot—almost as though shot by an invisible poison dart.

More or less typical was the young naval officer, a twenty-three-year-old ensign, who had been one of the first to arrive at Mass General. He had walked into the accident room under his own power, even briskly. His hands were badly burned and there were burns on his face and neck, but otherwise, aside from agitation and a flushed appearance, he seemed fit enough to be held aside as the flood of more desperate-looking victims began to pour in. Instructed to lie down and calm himself until he could be attended to, the ensign did not remain still long. He was soon up, pacing back and forth, waving his hands with the pain, then jumping frenziedly up and down. The next minute, he keeled over, face cherry red, scarcely breathing. It was found that is nostrils were deeply burned, the exposed mucous membranes layered with soot, and chest rales confirmed that he had inhaled quantities of destructive fumes. He developed alarming edema, or swelling, of the throat and then, rapidly began to experience obstruction of the upper respiratory passages, which an emergency tracheotomy failed to alleviate. Hours later he was dead.

There was a disconcerting number of others like him—too many. Postmortem examinations began to reveal a pattern the doctors found bizarre: massive pulmonary edema, which must have developed late and so suddenly as to be unaccountable; bronchial tubes choked with phlegm; a strange yellow cast to the mucus. Together, these symptoms improbably suggested the effects of poison gases—the deadly phosgene or mustard gases of the First World War, or of excessive inhalation of nitrous oxide, the so-called laughing gas used by dentists as an anesthetic, harmless in controlled dosages but extremely toxic otherwise. (And this seemed to fit the accounts of a number of victims who were able to recall having turned "giddy" during the fire, or gone "blank" to the extent that some of the most severely injured were not even aware they'd been burned—as though they'd been anesthetized by the fumes.)

What could have released such poisonous vapors? The evocation of phosgene in particular recalled for some physicians the tragic 1929 fire at the Cleveland (Ohio) Clinic that killed 125. It had started in the X-ray film storage room and spread through the hospital through the ventilating system. The heat of the burning film had converted its chemical coating to phosgene. (One doctor there, after long duty treating victims, died abruptly of the belated effects of phosgene while driving home.)

But the Cocoanut Grove? Yet, hadn't that building once been a commercial film storehouse? And further, didn't places like the Grove use chemical refrigerants in their cooling systems, which if combined under heat with oxygen could produce a phosgene-like gas? As for the similarity with nitrous oxide, that had a cellulose base, and weren't some modern decorative materials—artificial silk, leatherette—made from such a base? For that matter, wasn't film also a cellulose product? (How had the film manufacturers disposed of their deadly product after the Cleveland fire? Had they perhaps sold the excess to furniture manufacturers to convert into artificial chair coverings?)

It was a mystery. But it raised important, possibly vital questions that needed answering.

Within seventy-two hours of the fire, thirty-two of the 132 survivors admitted to Boston City Hospital died—all primarily of pulmonary causes. One of the first was Dr. Gordon Bennett.

His mother and father, summoned from Swampscott on the North Shore, had been at his side since 4:00 A.M. Sunday.

The Bennetts had lost their only other son to a brain tumor a few years before; Gordon, strong and healthy, was all they had, their sole hope for the future. Now they watched helplessly as that strong young body, with scarcely a burn mark visible, clung desperately to life. He hardly seemed to breathe. The tracheotomy had not seemed to help. They did not understand. The doctors attending him, including his former medical-school classmates, knew *what* was happening to him—he was, in effect, "drowning" as his own internal secretions, produced by the trauma of inhaling superheated flames, flooded his lungs—but they didn't fully understand why, or what to do about it.

At last, late on Sunday, Gordon Bennett simply could not draw another breath.

His parents learned that Gordon's boyhood sweetheart and fiancée, Edith Ledbetter, had survived. Injured but in fair condition at Carney Hospital in Dorchester, she would tell them later that she believed their son had gotten out of the Cocoanut Grove but had returned to save her.

Mass General lost seven of its original thirty-nine surviving patients in the first three days. The fifth to die, on Monday afternoon, was Buck Jones, just twelve days shy of his fifty-first birthday. Though he had been burned, the burns had not caused his death. He was another of the strange pulmonary victims.

It was late Sunday before the hospital could track down Jones's wife of twenty-seven years, Odille, who was staying at their ranch near Susanville in remote northeastern California. As it happened, their daughter, Maxine, who was married to the actor Noah Beery, Jr., had been visiting her mother just then, and the two women set out at once in a desperate race against time. Driving eighty rugged miles through the night to Reno, Nevada, they boarded a commercial airliner Monday morning for the long flight east. It was a frustratingly slow trip by twin-engine plane with numerous refueling stops and changes of aircraft across the vast country.

They got only as far as Cheyenne, Wyoming. When the plane touched down there in midafternoon Mountain Time, a sad message awaited them from Boston: he was gone. There was no point in continuing the journey. They returned slowly to California, haste no longer important.

Had he been granted a foreknowledge of where and how he was destined to die—in an Eastern nightclub—Buck Jones probably would have snorted with embarrassment, if not without some pained appreciation of the irony. And had he read some of the obituaries to follow, he could only have squirmed the more: invariably the public eulogies, from the soppily sentimental to a few tinged with malice, noted that the cowboy hero even at the end had "died with his boots on"—a reference to reports that he had sacrificed himself to help save numbers of others, and was so badly burned that when firemen dragged his body from the Grove he was recognizable only because of the expensive, beautifully handcrafted Western boots he always wore.

Jones, in fact, had not been much like the popular image of the "movie star": not really a public person in the showy sense, much less an habitué of glittery nightclubs; rather, he was a family man, a homebody, above all a true outdoorsman—a "Westerner" through and through. In Buck Jones, filmgoers got pretty much what was magnified up on the screen: a hard-riding, true-blue straight-shooter.

He was a legitimate cowboy, too—no less for having originated in Indiana. Born Charles Gebhart in 1891, he was reared in Vincennes but left there at age sixteen to join the U.S. Cavalry. After several years' service, first along the Mexican border and then in the Philippines, where he was wounded in a skirmish with Moro guerrillas, he was mustered out in Texas in 1913, a sergeant.

A performing cowboy troupe, the Miller Brothers 101 Ranch Wild West Show, was in Texas at the time, and Jones hired on with them as a horse wrangler. (It's unclear just when Gebhart changed his name to Buck Jones. He legalized it in California in the mid-1930s.)

In 1914, while playing in New York at the old Madison Square Garden, Jones fell in love with a bright young newcomer to the show, Odille Osborne. Petite, vivacious, and a skilled rider, Odille—unbeknownst to the show's management or to Jones—then was only thirteen. When he did learn her age, it didn't matter. The following summer, during a regular afternoon performance in Lima, Ohio, the twenty-three-year-old former cavalry trooper and the lovely fourteen-year-old equestrienne were married—on horseback in the center ring. (On the license application Odille gave her age as twenty-one.)

The pair, performing together with a variety of horse shows and circuses over the next few years, gradually worked their way west, to California. When Odille found she was pregnant in 1918, they decided to quit knocking about. Jones, scouting work, ran into some former wrangler pals who told him of good money to be made by experienced horsemen in the burgeoning film business. He had no trouble getting jobs as an extra and stunt man in the many cowboy pictures then being made by the Fox and Selig Studios.

Before long, with his physical stature, rugged good looks, and riding ability, Jones was called upon to "double" for such Western stars as William S. Hart, Tom Mix, and William Farnum. Next he began to get bit parts, and soon demonstrated that indefinable screen "presence" that somehow makes one actor stand out from others. In 1920, Fox gave him his first leading role in a Western called *The Last Straw*.

Almost overnight, Buck Jones became a household name in America. He made picture after picture—as many as eight a year—and by the end of the 1920s was one of the most luminous, and wealthiest, stars in the Hollywood firmament. He and Odille (who had retired with the birth of their only child) seemed to have it all: financial security from sound real estate investments, a beautiful home outside Hollywood, a large ranch out in the country, and a boat on which they could lose themselves.

The only setback to their idyll came in 1929, when Jones

decided to take a "leave of absence" from filmmaking to set out on a cross-country tour with an ambitious and expensive Western show of his own, which he called the Buck Jones Wild West and Roundup Days. The venture was a disaster, folding once in Denver after only six weeks on the road, then, after reorganization and refinancing got the troupe as far east as Danville, Illinois, collapsing for good when the hired manager absconded one night with all the accumulated receipts. Jones managed, barely, to make good his losses, and he and Odille (once more his costar, along with his white film stallion Silver Buck) trekked dolefully back to California.

By then "talkies" were the new rage in moviemaking—and the downfall of many a silent-film star, especially leading men whose voices and mannerisms were found unadaptable to sound and the new techniques. Jones had no such difficulties. And in 1930, at age thirty-eight, he was back before the cameras more dashing than ever.

Jones made scores of features and serials over the next dozen years—almost all old-fashioned good guys/bad guys (white hat/black hat) Westerns. Most were of the "B" variety, quickly and cheaply made, but they drew huge followings and made piles of money, as Jones (along with comparable cowboy stars of the day such as Hoot Gibson, Ken Maynard, and William "Hopalong Cassidy" Boyd) acquired a whole new generation of young fans. It was only toward the close of the 1930s that his light began to dim somewhat, as the style of Westerns changed with the rising popularity of singing cowboy heroes like Gene Autry and Roy Rogers. (Jones publicly deplored this new vogue as "sissifying" the genre.)

Nonetheless, as late as 1942 he was still considered a first-rate film commodity. Although the frequency of his releases had declined from the boom years, he'd made two pictures in 1942 (*Down Texas Way* and *Dawn on the Great Divide*), and had a new long-term contract with Monogram, which planned a series of "mature" Buck Jones films to begin the following year.

Thus, having been prominent in films through more than

two decades, spanning two generations of young admirers, he was hardly "coming out of retirement"—as was erroneously reported at the time—when he set out on his publicized cross-country morale-building tour for the government.

But now he was gone, abruptly and shockingly. And in death another legend—befitting his most memorable portrayals—began to be spread about the larger-than-life screen hero, that, disregarding his own safety, he had valiantly gone to the aid of others before succumbing himself.

Those closest to the real man had no doubt that if Jones could have he surely would. And maybe he'd even made an effort. But in the black turmoil of the Cocoanut Grove that night no one had been sure of much; and in the end, when all accounts were tabulated, no one who'd got out could honestly testify that *he* had distinguished himself in the emergency.

Most probably, as with so many others, Buck Jones had just been in the wrong place at the wrong time and was as powerless as any to do anything but perish.

At the end, indeed, he died *without* his boots on.

Jones's remains would be sent by train to Los Angeles, where, on December 7, there would be a cremation and memorial service at a Hollywood mortuary. Many of his actor friends as well as the authentic cowboys he'd delighted in associating with would pay him tribute and, at the request of the widow, play and sing some of Buck's favorite Western tunes. His daughter, with whom he'd been especially close, would lay a wreath of roses on the urn inscribed *My Buddy*. Finally his ashes would be taken up in a friend's plane and strewn over the coastal waters of the Pacific, where he'd enjoyed unwinding on his boat between pictures.

The day after the fire, a representative of the National Research Council's committee on burn research hurried to Boston by plane and offered Mass General (where Oliver Cope's research team had been working longest under a council grant) an exclusive opportunity to use a much-talked-about new antibiotic on its Grove patients: penicillin. Everyone in the medical field had heard of penicillin and its reputedly wondrous ability to fight bacterial infection, but so far its application had been limited to laboratory tests and controlled use on selected patients. The hospital jumped at the chance to be the first to try it in a general emergency, and the government set the wheels in motion.

Penicillin had been known since 1929. Discovered quite accidentally by an English bacteriologist who found a strange mold, or fungus, contaminating a culture of disease-related bacteria (staphylococcus), it had been isolated by Sir Alexander Fleming. The mold was found to produce a substance that was potently destructive of many of the microorganisms that commonly infect humans; moreover, it was unique in that its antiseptic effects were not toxic to human cells, which was a major drawback of the antibacterial agents then in use. Over the ensuing decade, Oxford University scientists, headed by Sir Howard Florey, had succeeded in concentrating and purifying the drug in liquid form, and had impressively demonstrated its curative powers, first in experiments on laboratory mice and then on individual humans with staphylococcal infections.

Pharmaceutical manufacturers in war-torn Britain had begun looking into penicillin, and in 1941 Florey had brought his findings to the United States, where interest was also quick to develop. By 1942—the two nations together at last in the war—British and American scientists were cooperating to rush the drug into major production for military use. As the year drew to a close, however, a viable process for turning out large amounts of the "wonder" serum had not yet been developed, and it was still in very short supply. The United States having been chosen as the key manufacturing center, most available raw components were then stored at the Merck & Company laboratories at Rahway, New Jersey, under lock and key and armed guard—quite, in effect, a military secret.

The National Research Council ordered a limited allotment from Merck for immediate shipment to Boston. Even working round round the clock, however, it took Merck chemists more than two days to prepare a supply adequate for clinical use. Then, late on Tuesday, December 1, the priceless consignment left Rahway by road caravan under state police escort, to be picked up by like escorts in each of the three other states through which it would pass. The journey to Boston took all night in constant heavy rains.

On Wednesday the 2nd, the fifth day after the fire, Cocoanut Grove survivors at Mass General became the first recorded general patients to receive dosages of penicillin (its use still so new to doctors that initially they administered it externally rather than intravenously or orally).

It was too late, of course, for Buck Jones and the six other Grove victims who had already died. But it might prove invaluable—if it was all it was cracked up to be—to those whose lives were still in danger. And if not—well, surely nothing would be lost.

chapter 11

Days after the fire, Boston—and much of the nation beyond—was still in a state of shock. The projected Cocoanut Grove death toll had reached the 480s and, it seemed, might possibly go as high as five hundred. Whatever the final number, it was the second worst such calamity in United States history.*

Boston, and Massachusetts, naturally had suffered by far the greatest number of victims. But the total casualty list, of both surviving and dead, included persons from twenty-five of the forty-eight states and the District of Columbia, plus one from Brazil.

Among the dead were fifty-one servicemen (including Marine Captain Walter Goodpasture of Columbia, South Carolina, whose wife, Mary, died with him); twenty-six others were hospitalized.

One soldier, Air Corps Private Harry Fitzgerald of nearby Wilmington, home for Thanksgiving after basic training, had

---

* The most destructive fire ever, in terms of fatalities, took place December 30, 1903, at the Iroquois Theater in Chicago. Of a standing-room-only audience of some 1,900 attending a holiday matinee performance of *Mr. Bluebeard* starring Eddie Foy, more than six hundred—many of them children—lost their lives. (The theater had billed itself as "absolutely fireproof.")

gone to the Grove with his three brothers for a last party before he went overseas. All four perished.

Perhaps the most symbolically poignant representation of the tragedy was a photograph published the next day by the *Globe* of an unidentified couple seated at a table, arms about each other, beaming into the camera, she wearing a corsage and he a boutonniere—apparently newlyweds. The next day the picture ran again, the pair now identified: they were John and Claudia O'Neil, who had been married in Cambridge just three hours before that photo was taken by the club photographer—the last ever taken in the Cocoanut Grove. The O'Neils had stayed to wait for the print, and they had died, and so had their best man and bridesmaid, who were with them. (The photographer, who had left after snapping that picture to take her negatives to a nearby photo lab, never made it back to the club. She was returning, a block away, when the Grove burst into flame before her eyes.)

If there was any saving grace to be derived from all the destruction, it was that so many had managed to *escape* death. Early estimates indicated that there had to have been more than a thousand people in all parts of the Grove when the fire struck. Thus, at least as many had got out as had died. (Of the estimated five hundred or more survivors, more than two hundred had been hospitalized. It was not then possible to gauge how many more might have sought private medical treatment, or how many lucky ones had not been injured at all.)

Considering the losses, it began to seem almost miraculous that so many had been saved. Or so, at least, it was being termed in official circles—where, of course, ultimate responsibility lay. But it would take a lot of rationalization and the long healing of time, before—if ever—a sorely wounded, personally aggrieved populace could come to see any silver lining in this affair.

It wasn't long before the public mood began to change— evolving from the paralysis of overbearing shock and misery to brooding, then anger, which festered into outrage, and at last to strident demands for retribution. *Who* was responsible for this inexcusable, wanton slaughter? The citizenry began to seethe like a great amorphous lynch mob. Somebody had to pay!

Clergymen sensed the ominous mood and pleaded from their pulpits for moderation, tolerance, even charity toward those who might ultimately be found at fault:

A priest: "Enter not into judgment with Thy servant, O Lord; for in Thy sight shall no man be justified unless through Thee he find pardon for all his sins. Let not the hand of Thy justice be heavy upon those whom the earnest prayer of Christian faith commendeth unto Thee, but by the help of Thy grace may they escape the judgment of vengeance who, while they were living were marked with the sign of the Holy Trinity...."

A minister: "In behalf of suffering souls we lift humble and awe-filled hearts to Thee, our Father.... Stunned by horror and pain and loss, we reach heavenward in contrition and faith that comfort and hope may enter our souls. In the still silence succeeding the shock we know a vacantness which only Thou canst fill.... Enable us, therefore, to rebel not at Thy providences. Shalt Thou send good and not evil?... Let Thy heart of compassion provide out of abundant reservoirs of wisdom and grace healing for our bruised souls and broken bodies. We believe Thou hast spared us for a purpose and will henceforth seek to realize Thy will...."

A rabbi: "Eternal, our hearts are heavy with the weight of human sorrow. May this tragic moment in our community teach us how fleeting is all of our existence.... O God, make Catholic, Protestant and Jew as united in life as they have been tragically united in death. Teach us to utilize the sympathy and righteous indignation of this moment to build a nobler city cleansed of corruption, greed and injustice. Make us wise, O Lord, in this hour of civic wrath so that we in turn commit no injustices. Purify our hearts of all hatred, of all malice, of all uncharitableness. Send Thy healing balm to all suffering hearts. Help the bereaved to feel that perchance in Thine inscrutable Providence Thou hast spared their dear ones vaster sorrow in the hard years to come. 'The Lord has given—the Lord has taken. Blessed be the name of the Lord....'"

The rabbi's exhortation seemed to reflect the people's turmoil more realistically than all the ponderously righteous oratory of public officials and the alternately somber or fiery editorials in the local press. If there be blame, he prayed, may those culpable indeed be revealed and purged, both for the sake of true justice and for the future security of the community; but above all—for God's sake—let some permanent *good* be forged out of this awful crucible of suffering.

But this was a result more easily hoped for than realized. There were so many questions to be answered. Could the fire have been purposely set? Or was it just a dreadful accident? In either case, had *criminal negligence* caused so many to be trapped by the blaze? At whose door—or doors—was blame justly to be laid? Who was prepared to deduce and pronounce such cosmic judgment? And—the stickiest question of all— who would be prepared, or willing to *accept* the responsibility?

Before the tragic weekend was over, several official inquiries were already underway. Mayor Maurice Tobin and Governor Leverett Saltonstall each had ordered immediate investigations to determine the cause and precise nature of so explosive a fire, and bring criminal charges before a grand jury.

Just about everyone in authority promised swift and rigorous action. State Attorney General Robert Bushnell and Suffolk County District Attorney William Foley jointly announced a criminal investigation by a hand-picked board of inquiry under the auspices of the Department of Public Safety. State Fire Marshal Stephen Garrity instructed Boston Fire Commissioner William Reilly to begin public hearings at once, inviting or, if necessary, subpoenaing testimony from all available participants and witnesses to the event as well as from all those in any way connected with the Cocoanut Grove. And meanwhile, Governor Saltonstall moved summarily to suspend the entertainment licenses of 1,161 other Boston establishments—cafés, clubs, hotel dining/show rooms, even taverns—pending their complete reinspection by fire marshals.

An obvious primary target—of both the official inquiries and the aroused citizenry—was the management of the Grove, specifically its longtime owner and operator Barnet Welansky. But Welansky was still confined to Massachusetts General Hospital in uncertain health.

Clouds of doubt also hung thickly over the city's own Fire Department and Building Department, whose inspection procedures and enforcement of existing fire laws appeared to leave much to be desired.

Among the first officially embarrassing, if not absolutely damning, pieces of evidence to come to light—a matter of public record and therefore easily accessible—was the last recorded safety inspection of the Grove.

It had been made and the report submitted just eight days before the holocaust, on November 20,* by a fire lieutenant named Frank Linney. In it the inspector noted: "A new addition has been added on the Broadway side used as a cocktail lounge, seating 100 people. No inflammable decoration. Main dining room seating 400 people...." Then, after summarizing other physical specifications of the premises, the report concluded tersely: "Sufficient number of exits. Sufficient number of extinguishers. Condition—Good."

On the heels of this, moveover, came the disclosure that two weeks prior to *that* report Fire Superintendent Bernard Whelan had formally warned the Grove management by letter that certain wiring for light fixtures in the club had been found to be lacking Fire Department certification because it was

---

* Ironically, the report had been misdated November 30—two days after the fire.

installed by "someone unknown to this Department." This was a violation of the existing fire code (which dated to 1908) and therefore "illegal. A permit must be obtained without delay," Whelan had advised. But there was no record that the Grove had complied in the interim.

These small admissions of possible official neglect or melfeasance were, however, less than entirely satisfactory. Even if a whole Pandora's box of bureaucratic bungling were eventually to be exposed, no one who knew Boston politics doubted the adroit backtracking and fancy footwork that was sure to be exhibited. Even if some heads *did* roll, there was little reason to expect they would fall from on high—only the most expendable would be sacrificed.

A spectacle of red-faced, and dragged-out, political finger-pointing could not, in any case, satisfy the demand by public and press for more tangible and immediate culprits to focus their outrage upon.

And, as it happened, there was such a one—a ready-made, uniquely vulnerable "culprit."

There was no disputing that the fire had broken out in the Melody Lounge. And several survivors of that room had told the media—most bitterly one Maurice Levy, whose wife and two companions were among the dead—that they had personally observed a bar boy strike a match close to a fake palm tree, and had watched the fire climb through that tree only moments later. Because of that single irresponsible act, they accused, all those hundreds of unsuspecting people had been incinerated! (What's more, the media had lost no time in ascertaining, the offending bar boy himself had escaped practically unharmed.) Clearly, they had found their culprit.

The young fellow in question was Stanley Tomaszewski, a student from Dorchester and, for the past few weekends, a part-time Grove employee.

Whatever he was thought to have done, the fact was that Stanley was far from irresponsible or careless or dumb; indeed, he was much closer to the classic personification of

Americana à la Horatio Alger. Not yet seventeen, he was a consistent B-plus student in the graduating class at Roxbury Memorial High School. Sober, solid, clean cut, Stanley, moreover, was a star athlete, a guard on the varsity football team who, every game, played sixty minutes on both defense and offense. He had a good chance of earning a full academic scholarship to Boston College but his aim was to enlist in the Army Air Corps as a pilot trainee upon turning eighteen. His older brother was already serving in the military.

The second son of struggling Polish immigrants, Stanley since late childhood had worked at odd jobs after school and on weekends—selling newspapers, delivering groceries, pumping gas—to help support the family. His father was a blacksmith, a trade for which there was no longer much demand, who had not had regular employment for the past nine years and who eked out a subsistence income as a local handyman; his mother had taken in laundry, sometimes hiring out as a seamstress or cook, but lately she had been frequently bedridden with gallstones and high blood pressure. With the elder son away, a heavier burden had been placed on Stanley. And he had responded as always: at the end of the football season, in addition to his other after-school jobs, he'd started to work Friday and Saturday nights at the Grove. His classmate and best friend Joe Tranfaglia, a nephew of maître d' Angelo Lippi, already worked there. Yet with all this, Stanley had managed to keep up his grades; and he dreamed of a better future....

When the fire started in the tree, then swiftly spread to the billowing tent ceiling, Stanley battled it single-handed at first, slapping at it with bar towels and splashing it with water and seltzer. Then, when nothing he or the other employees could do seemed to have any effect, he'd tried to pull down the tree and the burning material overhead and, for his trouble, got burned by a shower of fragments. That was when bartender John Bradley shouted to Stanley to direct people out through the door to the kitchen. He and Bradley herded together as many as they could, but the place was in an uproar by then and few had listened or understood. Finally they had managed to lead two small groups out of the inferno and safely into the alley behind the club.

Out on the street, shaken, dazed, Stanley took a few minutes to catch his breath and try to compose himself. Glassy-eyed, he watched as the blaze, with incredible speed and fury, seemed to devour the whole club, section by section. Then he thought of his friend Joe Tranfaglia. Terrified, he circled the block again and again, sidestepping massing firemen and their equipment, shouldering past screaming onlookers and hysterical escapees, eyes straining for some sight of Joe. At one point, at the corner of Piedmont and Broadway, he came upon a young woman he recognized as an employee of the club and asked in desperation if she'd seen Joe Tranfaglia; but she, as frightened as he, said she didn't know him. A fire officer, wearing a peaked cap instead of a helmet, came over then, having spotted Stanley's white bar jacket, and asked if either of them had any information on how the fire had started. Stanley, mind in a turmoil, just shook his head dumbly and moved off again to search for Joe.

He wandered the streets for hours. It was 4:30 A.M. before he made his way home to Dorchester. In their mean tenement flat, his parents, sick with fear for him, were waiting up. Foggily Stanley calmed them. They bathed his scorched hands, and then he fell into bed exhausted.

He slept only a few hours. Awakening early Sunday, he quickly dressed and went out to a pay telephone to call Joe's house. The Tranfaglias were near their wits' end: they'd still had no word from their son. In the icy grip of fear, Stanley took a trolley up to Boston City Hospital, where the newspapers said most of the fire victims had been taken. The hospital looked like a war zone, but he managed to find a desk staffed by women who had long lists of all those who had been brought in. Joseph Tranfaglia was not listed. Stanley then rode across town to Mass General. The answer there was the same. Bewildered, but with the tiniest spark of hope, he returned to Dorchester.

His mother and father were grim, their eyes red-rimmed.

There was a message from the Tranfaglias: Joe was dead. His body was in the Southern Mortuary.

Numb, Stanley thought about what he should do. He relived the fire, where it had begun, *how* it seemed to have begun. Was it possible that *he*—? The thought had not crossed his mind until that moment. The match. But it had not touched anything, he was sure of that. And he had blown it out, and crushed it underfoot as well. He had not seen the flames at once—they had come seconds later, maybe as long as a minute. He had no idea how the fire had started, but he was convinced it was not because of anything *he* had done. All the same, he'd been there, right in the middle of everything, when it did break out, and it occurred to him that he really ought to tell the authorities whatever he knew.

So Stanley left his house again early Sunday afternoon and went back uptown, to the Boston Police Headquarters on Berkeley Street, only blocks from the gutted hulk of the Grove. He walked in, identified himself to the desk officer, and said he wanted to give information about the fire. The policeman listened to him a few minutes, his gaze narrowing at the pale, dissheveled youth with the hollow eyes, unshaven chin, and flat voice, then asked him to wait while he rang upstairs. After speaking with the watch commander, he told Stanley to go home but to be prepared for a summons to appear before a formal hearing conducted by the fire commissioner.

When Stanley left, the desk officer called the press room and told the reporters there about his visit.

The next afternoon, Stanley had just come home from school when a clamor arose and grew in the street outside the Tomaszewskis' building. From the window, he could see a sudden gathering of people below. They were glaring up, some waving fists and some shouting, seemingly angry although their actual words could not be made out through the closed window.

Moments later there was a fierce banging at the apartment door. Startled, Mrs. Tomaszewski opened it a crack. A dozen or more intense-looking men were gathered in the

141

hallway. She asked what they wanted, what was going on.

"We want Stanley Tomazooski!" they demanded.

(The correct Polish pronunciation is TomaSHEFski.)

"There is no one here called that!" she cried and slammed and locked the door.

They pounded again, calling the Tomaszewskis out, swearing viciously. Mr. Tomaszewski ran from the back room with several lengths of two-by-four, which he hastily jammed against the door.

"Tomazooski, Tomazooski!" they chanted, almost in cadence with their pounding.

Stanley came out of his room in amazement, and his mother urgently shushed him back inside. He retreated, still not understanding; but hearing those harsh voices, the threatened violence, he became truly scared for the first time.

But soon, approaching sirens drowned out the commotion below, and then squads of policemen (responding to the calls of anxious neighbors) were scattering the raucous intruders. A couple of officers came to the flat and they decided it might be best, for the present, if Stanley went along with them.

There were still knots of muttering vigilantes down the street, being kept a distance from the house by officers brandishing stout billy clubs. "Murderer!" one bellowed when Stanley emerged, and others took up the cry, heaving against the police line. "You'll get yours, murderer!"

Unnerved, disoriented by the nightmarish experience, Stanley allowed himself to be hustled into the back seat of a police cruiser and sped away.

When the police detail reported these events to their superiors, a serious conference was held by the department's top commanders—including Commissioner Joseph Timulty himself. The young man Tomaszewski, from all reports, could be a key witness in the inquiry and they could not risk any premature harm coming to him; but all else aside, he was a citizen being threatened with mob violence and was due all reasonable police protection. It was decided that they would keep him in protective custody indefinitely, at least until they could be certain he was no longer in personal danger.

A room was discreetly booked at the Kenmore Hotel near Fenway Park. And that evening, after dark, the high school senior was escorted there by plainclothes officers. Two men were to stay with him at all times, around the clock.

They would be Stanley Tomaszewski's only companions, and that hotel room his home, for the next two months.

chapter 12

Excerpt from an editorial in the *Christian Science Monitor*, the august national daily published from Boston, some days into the Fire Department's public hearings:

> Naturally no one wants to take the awful responsibility for the killing.... But the process of whitewashing is sickening. If the results of investigations so far were to be accepted the blame would rest wholly on the crowd which was panicky or befuddled, on a sixteen-year-old bus boy who may have accidentally touched an artificial palm tree with a match, and on Boston's building laws. The law is a particularly tempting and undefended scapegoat.
> ... Particularly remarkable is the record of the city officials charged with inspection of such premises. The Building Commissioner maintains that the place had proper and adequate exits, pointing out that the law lists nightclubs as restaurants and does not impose the same requirements on them that it does on hotels and places of assembly like theatres and churches. This legal deficiency can and undoubtedly will be made good, not only in Boston but wherever it exists in the Nation.
> But one reason it exists is because dine-dance-drink places have too often been closely connected with poli-

ticians. For this same reason, it will be very hard to convince the public that proper enforcement of the laws already existing would not have prevented this disaster. How much good are more laws if the report of a fire inspector made on November 20 could describe the Cocoanut Grove conditions as "good"? Eyewitnesses testified that this was a "flash" fire, spreading with unbelievable rapidity in "stuff that looked like straw" on the ceilings and through fake palm fronds. Yet the fire inspector says that he tried to light several palms with matches and his report says "no inflammable decorations."

This isn't good enough. No one was to blame! But action will be taken to prevent another Cocoanut Grove, and somebody could have taken action to prevent this one. And not merely by passing new laws. A national expert on fire prevention... says it is too soon to determine just what part was played by "the chaotic condition of Boston's building laws, incompetent enforcement, political influence and careless management"....

All these are causes which can be dealt with. All could be improved were this fire to purify Boston's concepts of city government. For a generation Boston has been satisfied with government on the nightclub and racetrack level. The laws will be improved. But conditions under them will be better only as citizens insist on better men to enforce them.

The public hearings, conducted personally by Boston Fire Commissioner William Reilly, had begun at Fire Department Headquarters less than twenty-four hours after the fire. With the commissioner were representatives of various municipal and state agencies as well as of the federal government, including Army and Navy brass and agents of the FBI.

The other official investigations were not accessible to the public: Fire Marshal Garrity's separate inquiry for the Commonwealth and the joint criminal probe of Attorney General Bushnell and D.A. Foley. There was some urgency to the latter investigation—and also, therefore, to Garrity's, whose findings were expected to bolster any evidence for a criminal case—because the term of the grand jury currently sitting would expire on December 31, and the lengthy delay that would be caused by waiting for a newly empaneled jury was unpalatable to all involved. Immediate action was required, not only to serve justice, but politically as well.

But Reilly's open forum *was* mainly for public consumption. And whether or not the information it brought out would prove ultimately to be revealing or instructive, it did, with its parade of public and private witnesses—those who blinked and squirmed in the harsh glare of critical scrutiny, and those who came wretchedly to affirm the horror and grief—capture the public attention and receive daily front-page coverage in the press. The hearings would continue six days a week, Sundays and holidays excepted, for the next seven weeks.

Among the first witnesses called by Commissioner Reilly was one of his own inspectors, the lieutenant, Frank Linney, who, the week before the fire, had turned in the routine report terming the Grove's safety conditions "good." Linney, an erect, gray-haired, thirty-year veteran of the department (with two commendations for heroism), now was an embarrassment; but there was no choice but to air his seemingly colossal gaffe, and see whether any redeeming information could be wrung from it.

Linney was pressed to elaborate on his written report, whose official terminology had been so economically undetailed about fire regulations—specifically with respect to his one key (and now glaring) phrase: *No inflammable decoration*. The abashed lieutenant testified that in fact as part of his inspection on that date he had tested in turn, with lit matches, the leaves and bark layers of the fake palm trees, the satin ceiling material, the leatherette and rattan and netted wall coverings, even the upholstery of some of the furnishings, and none showed the slightest susceptibility to combustion. He added that as the management of the Grove was on record as having flameproofed all such suspect decorative trappings throughout the club, he had found no reason to believe otherwise.

Had he, then, no clue as to what could have touched off the blaze, to say nothing of why it should have grown with such speed and fury? From his own personal knowledge, Linney said, none. But could he still maintain, in view of what did happen, that as of November 20 the Cocoanut Grove had merited a "good" rating in regard to fire safety? Linney flexed his jaw muscles and in a hoarse but positive voice answered: "Yes, in my opinion...."

A meteorologist from Harvard came forward with testimony that both lent some support to Linney's stubborn assessment and also somehow damned it. The weather scientist said that the day the inspection was made happened to have been the wettest in the entire month of November. The general dampness, he maintained, plus the fact that Linney's match tests largely had been made in the New Lounge, still not quite finished and partially exposed to the elements by workmen coming and going, could well have inhibited casual combustion. (Laboratory tests subsequently corroborated this thesis: on the next day with conditions similar to those recorded on November 20, samples of the same combustible materials used in the Grove did not catch fire readily when matches were lit and held to them; on the next day of markedly low humidity, however, they tended to ignite more quickly.) The meteorologist went on to criticize the usual mode of fire inspection for ignoring, or ignorance of, the factor of relative humidity when assessing hazard. From all experience with such enclosed, overheated, and underventilated environments as are common to nightclubs, it seemed a reasonable estimate that relative humidity in the Cocoanut Grove any night of full operation would have been no higher than 12 percent; and that, he said sternly, was rated by forestry specialists, who had serious cause to monitor such pertinent phenomena, as a most dangerous fire threat—tantamount to a tinderbox.

His testimony was, in sum, an indictment of basic and

hoary Fire Department rote. But it served to leave Frank Linney pretty much alone out there in the woods.

Next to come up was the matter of Fire Superintendent Whelan's earlier warning to the Grove that it was in violation of the fire code for unauthorized installation of electrical wiring and that a legal permit was required at once. Had this order been complied with prior to the fire? No one could say. There was no record...

Then it was learned from an inspector in the department's Wire Division that the Grove also had failed to secure a permit for electrical work done in the New Lounge. Two notices had been mailed, he said, and a third and final one was due to go out December 1, after which, if there was still no response, the club's meter would have been removed and its electricity cut off.

Two electricians who had done recent work at the Grove were traced and summoned before the inquiry. One was a twenty-six-year-old shipfitter at the Boston Navy Yard named Raymond Baer, who said he had performed odd electrical repair jobs for the club in his spare time over the past several years and had installed wiring both in the Melody Lounge and, more recently, in the New Lounge. He was not a licensed electrician, had never had a license, and so had not applied for the installation permit required for the new room. He supposed it was his work that had been tabbed "unauthorized" and "illegal," but he felt he had done a professional job nonetheless. He said that he always tried to follow the code.

In the Melody Lounge, Baer said, he had put in some ceiling fixtures and also done some work on the motor that powered the revolving stage (for Goody Goodelle's piano) inside the bar. How about the lights in the palm trees, he was asked, had he wired them? With some hesitation, Baer replied: "That... I don't remember about that."

Didn't he understand he was violating the law by doing such work not only without an approved license but without a permit for the specific job?

Baer said that when he was working in the Melody Lounge the designer of the club, Reuben Bodenhorn, and the contractor who was building the New Lounge, Samuel Rudnick, "were around," and that he had gotten the impression from them that a permit was being obtained. As for the New Lounge, which he'd wired during October and November, Baer recalled that the owner himself, Barnet Welansky, in fact had mentioned to him having received notice from the Fire Department about a permit; here, too, he said, Welansky had seemed to indicate he need not concern himself personally.

Finally Baer stated that all the wiring work was not completed when the new room opened about a week before the fire; lighting had yet to be installed in the basement, and current for lights on the second floor of the ramshackle building in which the New Lounge was housed had to be switched to a different meter. At that point, he said, he had called upon the Grove's former electrical maintenance man, Ben Elfman, to see about getting the permit. Baer said he had even gone to the club's bookkeeper, Rose Ponzi, to get $5 for Elfman to pick up the permit; and he understood that Elfman had spoken to some fire inspector or other.

Why call on Ben Elfman? "Because the pressure about a permit felt like it was getting tight, and Ben *is* legal."

Benjamin Elfman, of Roxbury, was a master electrician who, while employed by the Grove, had taken on young Baer as a part-time helper and who had since gone into business for himself. He testified that during Thanksgiving week Baer had asked him to come in and appraise some tricky electrical work that needed to be done on the floor above the New Lounge (by then in operation). He and Baer had looked it over together, and Elfman saw that the wiring up there was very old and much of it needed replacement. Baer asked him if he could get a permit for the job, and Elfman said yes but he wanted to check first with Welansky to see if he definitely should proceed with the intricate job. He learned that Welansky was in the hospital, but someone else at the Grove— he couldn't remember just who, but it had to have been somebody with the authority—told him to go ahead. He said he'd then "contacted" an officer of the Fire Department's Wire Division to arrange for the permit... but, of course, there'd never been a chance to start on that wiring job.

Elfman was asked whether, in first seeking the management's express approval to proceed, he'd had some reason to think they might have wanted the job done *without* the formality of applying for a permit.

Not necessarily in this case, Elfman answered. "But I have seen a lot of jobs in my time where the work *was* done without a permit."

"Electrical work done in Boston without a permit?" Commissioner Reilly demanded, red-faced.

"Yes, sir," responded Elfman unhesitantly.

Next, a neon-sign supplier named Henry Weene was brought in. Weene had been contracted by Barney Welansky back in October to install neon lighting in the New Lounge while it was under construction. He was asked point-blank if *his* work had been performed legally—that is, with the proper permit.

Weene seemed more than usually discomfited in the spotlight. "When Welansky told me what he wanted," he replied edgily, "I said the job needed a master electrician and a permit, and either I'd make the arrangements or he could if he knew somebody. But he told me it wasn't necessary. He said we wouldn't have to go through those formalities, because—" here Weene flushed and his eyes darted about as though seeking help from somewhere; then he swallowed and blurted, "—because he and Tobin fit. That's what he said. 'They owe me plenty downtown,' he said."

Mayor Tobin?

Yes. That's who Weene had taken Welansky to mean.

Naturally, this produced rather a stir. Reporters scurried to City Hall to confront Mayor Maurice Tobin, who (having only a day or so before delivered a widely heard and reported radio address movingly entreating citizens to temperance in their demands for swift justice in the Cocoanut Grove affair) now fired off an outraged denial of Weene's allegation: "Since I have held the office of mayor, I have never—wittingly or unwittingly—permitted any winking at the law, or any violation of the law!" Either Weene was lying or he had misunderstood, intimated Tobin, or if Mr. Welansky *had* said such a thing, he'd had no cause or right to. "Barnet Welansky may not assume his relations with City Hall are such," the mayor sniffed, "as to permit scoffing at any aspect of the law." (The fact that only the previous summer Mayor Tobin had appointed Welansky a member of Boston's top-level War Rationing Board was duly recalled by the press as possibly relevant in this context.)

Thus began to emerge—or to be substantiated, for many had long suspected it—the ill-defined but still recognizably seamy pattern of favor, not to say license, enjoyed by Barney Welansky's Cocoanut Grove (and how many other places like it?) among the city's public offices, and even with high-ranking individual administrators. Whether or not what Henry Weene said was accurate, the dice had been rolled and the point established.

Official after official called before Reilly's inquiry, representing Building, Health, and Police departments (the agencies supposedly responsible for regular inspection and critical evaluation of public establishments for fire and other health or safety hazards), could only concede, some defensively, others truculently, that there was indeed no record of the Cocoanut Grove's ever having been denied issuance of permits for anything—from licenses to serve liquor and food to certifications of building maintenance and fire safety standards—at least not since the Welanskys had taken over back in 1933, shortly after Repeal.

A member of the city Licensing Board confirmed that the Grove's operating license seemed to have been perennially granted without formal review or even a hearing. He read from the club's 1942 license, as usual approved routinely, numbering its furnishings as one hundred tables, four hundred chairs, thirty barstools. Then he produced a copy of the recently submitted renewal application for 1943—which the board had not yet got around to passing on—requesting the addition of thirty seats, presumably for the New Lounge. Yet it was apparent from testimony of patrons and some employees that those thirty new seats had already been in place on the night of the fire, before the board had ruled yea or nay. Moreover, by any count the stipulated number of barstools, thirty, was obviously far out of line with reality: in fact, the hearing found, there had been some eighty in use at the Caricature Bar alone, with another two dozen or so in reserve at a small back bar used for overflow; about twenty-five in the New Lounge; and another seventy-five to eighty downstairs at the Melody Lounge bar—more than two hundred stools in all!

Survivors of the fire had declared without exception, many now with bitterness, that the Grove unquestionably had been overcrowded that night, even grossly so; and this was corroborated, not without reluctance, by a few who'd worked there and who indicated it may well have been the most crowded it had ever been. The great majority of employees who testified—notably musicians and entertainers—steadfastly refused to go on record as saying the place had been any more crowded than usual. "It was busy, but no more than on any Saturday night," most insisted.*

Primarily responsible for ruling on occupancy, safe capacity, and proper egress facilities was the Building Department, which was beginning to seem the most deficient of all. Commissioner James Mooney and his chief inspector of the department's Fire Prevention Bureau, Theodore Eldracher, appeared before the hearing with a selection of charts and sketches purporting to demonstrate that the Grove's layout

---

* Later, however, and privately, several of these—especially the musicians, who had played many nightclubs before and since and who, after all, had been sitting on the bandstand watching the club room fill up—would agree they had never seen the Grove so crowded.

and placement of exits should have allowed, on paper at least, for the escape of no fewer than 1,300 occupants without great difficulty. Eldracher said further that on his own last inspection of the premises, during the recent construction of the new addition, the only citation he'd found cause to give was for the lack of a metal fire door in the corridor between the New Lounge and the main dining room. (He'd been given to understand that the club had ordered such a door but, Eldracher added ruefully, that delivery had been held up because of wartime restrictions.) Therefore, the two were asked, the new room had been permitted to open for business without ever having been finally approved by the Building Department? Evidently, they murmured. Well, why? They would have to look into that.

But, such technical oversights aside, how could 1,300 people have been expected to get out quickly in an emergency when only two marked exits in the entire place were operative? And both of those, as it turned out, were defective from a safety standpoint: the revolving door at the main entrance was not designed to accommodate a mass flow in a hurry; and of the doors at the Broadway entrance of the New Lounge, the inner one opened only *inward*, creating a natural barrier to those trying to push through.

How did they explain, or rationalize, or *justify*, all the locked doors, the concealment of exits and other potential escape routes (such as windows)? *How* could such a mass of people have escaped if they couldn't get out where they should have been able to in the first place, and were unaware of the locations of other outlets as well?

Mooney and Eldracher sat glumly. There was no easy answer. There *were* some violations, it was now clear. But at the same time some of these actually were permissible under existing law! The *law* was basically at fault—pitifully outmoded, not geared to modern conditions, in need of wholesale updating. As it was now, they often felt their hands were tied....

* * *

The police officer who for more than three years had patrolled the beat that included the Cocoanut Grove and other nightclubs in the area described how he had watched in astonishment and horror from just a block away as roiling flames and smoke "just like that" engulfed the club, and how he had then sprinted over to help pull people out. Asked specifically if in his time on that beat he'd ever had occasion to file a report of any sort of violation in connection with the Grove, he said no. He'd never had any trouble around there, and he'd never been ordered by a superior to inspect the premises for any reason. The only times he'd actually set foot in the place were to deliver license renewals.

What sort of occasion, had there been one, might have warranted an official report? "I suppose some kind of arrest situation—an altercation, you know, like a fight, or a complaint from the club that somebody was trying to beat his check, or maybe from some customer. Or in case of fire..." He had reported *this* fire, then?

"Well, no," the officer said. "I knew the captain was in there, so I figured he'd take care of it."

The captain was Joseph Buccigross, district night commander, who on return from vacation had chosen that Saturday night to check out the Grove's New Lounge, which had opened while he was away. While it was part of his assignment to check regularly on the conduct of local clubs and bars—watching for illegal gambling, hookers on the make, liquor served to minors or drunks, and also dangerous overcrowding—his "inspection" that night really had been only semiofficial. Just back on the job, he'd wanted to get a feel of the new place. Nonetheless, he was on duty, though in plainclothes, and so, as was customary, before going in he'd informed the beat cop of his presence.

All this, or at least the salient facts, had already come out in the papers, and there had been hot mutterings of resentment expressed by many in Boston about the police captain's part (or non-part) in the disaster. Buccigross, along with Welansky and Byrne, had got out safely—among the *first* out, the bitter story circulated—with the initial rush of patrons from the blazing lounge. Why did *he* get out? Why had he run? What had be been doing there (and out of uniform) in the first place? Boozing on the city's time? Or worse, *collecting his payoff*? "Everybody" knew places like the Grove were hangouts for gangsters—maybe *especially* the Grove, which had once been run by a notorious mobster and been taken over by his own shyster lawyer, Welansky, when the guy got himself bumped off. A lot of people, in fact, were no less certain that Welansky himself was still connected with big-time criminal types, along with being "in" at City Hall and down the line. And sure as God made rotten apples, cops were on the take to look the other way in these places. So what was this nightclubbing captain doing while hundreds of innocent people were being slaughtered?

It was not an unreasonable question, as unfair as the emotional assumptions behind it may have been. Captain Buccigross appeared before Reilly's inquiry a shaken man; he was an experienced, tough cop, but all this furor had left him in a limbo of unnerving uncertainty. He gave his account flatly, describing how he'd been knocked off his feet in the spontaneous wild scramble for the exit, and, unable to resist the hysterical surge, found himself propelled out into the street. He'd tried to go back inside, he said, but... he had *not* been drinking, hadn't had even a soft drink. And he was *not* there for a payoff. He'd never done that sort of thing in all his years on the job.

That was all there was to it. He was sorry, but...

Hadn't he seen how exceptionally crowded the place was? Wasn't it part of his job to do something about that?

The lounge was pretty full, yes, said Buccigross, but he hadn't yet familiarized himself with the new setup, still hadn't gauged what its safe capacity should be. (As for the rest of the club, he hadn't looked inside that night; he would have done that next.) But in any case, he pointed out, it had happened so fast he didn't see what he could have done to change anything. Maybe if he'd just walked in, taken one look around, and said then and there, "Okay, there's too many people in here, close it down," Buccigross mused...but by the time he could have got anybody moving, and the way that fire had hit, *wham!*—would it have turned out any differently? The captain shook his head dolefully. There was just no way of telling, now....

Buccigross's top bosses, Police Commissioner Timulty and Superintendent Edward Fallon, came to the hearing and backed him all the way. They said it was his job, his *duty*, to spend time in places like the Grove; and further, in such situations it was preferred the men wear civilian clothes so as to be less conspicuous. (What Fallon refrained from mentioning, however, was that when he, himself, had arrived at the Grove during the height of the fire—with Buccigross then busily trying to help direct rescue efforts—he had pulled the captain out of the line, flayed him for being out of uniform, and ordered him back to his station to change.)

If, in the view that was formulating, the municipal establishment was being judged increasingly suspect (if not already guilty) of insensitivity and inefficiency at best, or, at worst, of dereliction and even corruption, there was almost no question in anyone's view as to the *source* of such evils: the shadowy recesses of the Cocoanut Grove, which catered to laxity and fed influence—and namely, the wheeler-dealer "shyster" lawyer who'd operated the place with apparent impunity all these years, Barnet Welansky.

Commissioner Reilly had tried time and again to get Welansky before his board of inquiry. But for more than a week after the fire and the start of the hearings Welansky had remained on the critical list at Mass General. A week later he'd been discharged, but by then his personal physician had signed an affidavit stating that Welansky required a minimum of two more weeks convalescing in seclusion at his Brighton home; he was extremely weak, and thus under no circumstances could he be allowed to testify at that time. (Nor could the panel avail itself of another key figure, one who'd been associated with the Grove even longer than Welansky, virtually from its inception in the 1920s: the suave maître d'hôtel Angelo Lippi, who was also still bedridden after many weeks.) So, if only as a timely symbol, Reilly settled for Jimmy Welansky, Barney's younger brother, who had been overseeing the club on the night of the fire.

Jimmy did his brother no service by appearing. A slight, pale, careful man, Jimmy was not able—or did not propose—to contribute a whit of enlightenment to the proceedings at hand. His responses to all questions about the club's operation were monosyllabic and noncommittal, nor would he volunteer so much as an educated guess. According to Jimmy, he knew nothing about how the Grove was run—notwithstanding the established fact that actually he'd had longer and broader experience in the restaurant business than Barney himself: currently he was not only the proprietor of the Rio Casino but also had an entrepreneurial interest in the Circle Lounge out near Boston College; and in the past, he had managed the Theatrical Club and even a hotel, the Metropolitan.

But all Jimmy would say was, Barney had *his* business and he his own. He had no particular concern with management of the Grove; since his brother's illness, now and again he would trouble to stop by, to see how things were going and to let the employees know somebody was around, but that was it—Barney had other people there who knew better how he wanted the place run. About certain exits that had been found locked, Jimmy did venture an opinion: he "imagined" that they may have been locked during or after the shows to prevent deadbeats from skipping with their checks unpaid, which, he said, was all too common. He couldn't say, however, if it had been regular practice at the Grove.

Tonelessly, without any apparent defensiveness, Jimmy maintained he just didn't know. Fire inspections, licenses and permits, compliance with building codes—he didn't know about any of that. Flameproofing of decorative materials? He was sure that must have been done; he "assumed" it had been... but he didn't really *know*.

By his steadfast reticence and seemingly ingenuous profession of ignorance of his brother's affairs, Jimmy—whatever his intent may have been—succeeded only in exasperating those conducting and attending the hearings and in convincing the public that there must have been *plenty* to cover up at the Grove, and that he just trying somehow to keep the heat off Barney.

And it was true, as far as that went; in his way Jimmy *had* been shielding his brother—or so Daniel Weiss understood as, awaiting his own turn as witness, he'd followed his uncle's uncommunicative testimony.

But in Weiss's eyes, the intent was not merely a clumsy or dishonest attempt to mislead. It was, first, a pure and simple matter of family loyalty, of one kinsman standing up to take the beating for another who was fallen; and the Welanskys were a fiercely devoted family. Jimmy would not say *anything* that could in any way contribute further to his brother's distress—whether or not he did know anything about the workings of the Grove. But that was the other side of the story: not only did Weiss know that Uncle Jimmy was nowhere near as dense as he'd made himself seem, but he suspected that neither had Jimmy been so evasive as he'd come off. From Weiss's own experience at the Grove, he felt it not unlikely that Jimmy did *not* know as much about the place as everybody supposed he must. Before Barney fell ill, Jimmy had only very occasionally come around; as he'd stated, he had his own businesses to keep after, and as close as they were he'd never interfered with his brother's; when he had dropped in, it was usually just social—What's new? How's everything going? Have a drink. But, of course, Weiss realized sadly as well, nobody would believe that. Nobody wanted to.

In any event, Jimmy's apparent obfuscation had a dual effect upon the investigation: it not only served to fortify, by its very negativism, the public conception of Barney Welansky's "guilt"; but it also opened yet another can of worms for both the brothers.

Exhumed now from musty Licensing Board files and old court records was evidence confirming that both Barney and Jimmy Welansky had been involved in the past with criminals and criminal or civil legal actions. Barney, as was known, had been counsel (with his then law partner, Herbert Callahan, who was one of Boston's top criminal lawyers) to the late Prohibition rackets boss, Charles "King" Solomon, when Solomon owned the Grove. Though ownership undisputedly had been Solomon's, never in the three years he had it—thanks to machinations by Barney Welansky?—was it registered in Solomon's own name. Then, following Solomon's violent demise, somehow, mysteriously—unexplained by city records—the club had fallen into the hands of the attorney, Welansky, through some kind of inexplicable arrangement (an underhanded power play? a swindle?) with the widow. And from that day, early in 1933, ownership of the Grove had shown up as a murky, tangled series of apparent dummy or proxy registrations—relatives of Welansky, or employees of the club or of his law office, but never, until only the past year, Barney Welansky himself. Why, people asked, so much corporate/legal hocus-pocus? (And why had the city never *demanded* more explicit information?)

It was further recalled that only a few years before, Welansky had been a defendant in an odd civil suit that had nothing directly to do with the Grove but that did involve—pertinently now, perhaps?—the city's building laws. He and a Boston fight promoter had subleased the old Lyric Theatre and secured permits to remodel it as a hall for playing beano (a low-stakes gambling game then in vogue, similar to bingo or lotto). The owners of the Lyric—who had originally leased it to a third party with the intent that it continue to be run as a theater—sued, and the beano operation was suspended. Stymied, Welansky and his partner had moved to turn the hall back into a theater. But the Building Department denied that plan because—and here was the muddling part of it—the Lyric's license as a theater had been nullified by its conversion to other use, and the law pertaining to second-class structures such as this prohibited its being rebuilt as a theater. Welansky and the other man were not only left holding an empty bag but were found by the court liable to the owners for the sum of $14,134.

As for Jimmy Welansky, he, too, had been the subject of notoriety during the 1930s—in a matter far more lurid than building code violations: the murder of a gambling associate of his in the lobby of the midtown hotel that Jimmy then managed. On December 17, 1937, David "Beano" Breen was shot to death at the Metropolitan Hotel. There were no reliable witnesses, or anyway none who came forward, and after a long inquest proved inconclusive the murder was recorded as "unsolved." The judge, however, in his summary of findings, issued a scathing commentary that left little doubt of his unofficial opinion: "There is evidence," he wrote, "that David Breen and James Welansky were engaged in a gambling enterprise at Nantasket [a beach resort south of Boston] during the summer at which $20,000 was lost, and that Welansky was left to stand the loss."

That sordid affair was brought to light again as reporters—and, no doubt, criminal investigators for the attorney general as well—sought to examine the police records of the Breen case to see whether they might contain any evidence relevant to Jimmy Welansky's qualifications to be issued a liquor license, which he'd evidently had no trouble obtaining. And here they got a surprise: at police headquarters it was found that those records, with Welansky's prints and mug shot, were *missing* from the file—with a notation that they had been "removed by order of the Police Commissioner"! Why? When?

Commissioner Timulty immediately denied having issued such an order and said moreover that he wasn't even aware the folder had been removed. Other police officials explained, or tried to, that they kept two sets of files at headquarters, one for ongoing or "active" cases and another for "inactive," ones where the records of those "found not guilty" were held; information about Jimmy Welansky relative to the Breen case undoubtedly was in the second file, they said, and would not be released to anyone except on official request in pursuit of a current criminal investigation. Which suggested that Bushnell's investigators and grand jury might get to see those records, but not the public.

But, the press argued, Jimmy Welansky had *not* been "found not guilty" in the Breen shooting. He had not been formally accused, but neither had there been an unequivocal determination of the case; therefore, since resolution *was* still pending, should not that murder be classified even yet as "active"?

But the police remained unmoved. Either way, the file was unavailable.

What was going on here? rose the public outcry. Was everything these Welanskys touched corruptible?

Even the Grove's innocuous little bookkeeper, Rose Ponzi, drew special attention and aroused speculation—not for anything she was found to have done but for who she was, or rather had been. She was the former wife of one of the most notorious con artists and swindlers in American history, Charles Ponzi, who in the 1920s had taken gullible investors for untold millions of dollars.

A destitute twenty-one-year-old Italian immigrant when he'd come off a boat in Boston early in the century, Ponzi had drifted around the United States and Canada for fifteen years, supporting himself with odd jobs (and an occasional scam, for which he did some jail time in both countries), before returning to Boston about 1918. There he met and married petite, reserved Rose Gnecco, daughter of a produce dealer in the North End, and took a job with his father-in-law's concern. At age thirty-seven he was a sixteen-dollar-a-week fruit and vegetable peddler with negligible prospects for advancement.

A year later, Ponzi was a multimillionaire, hailed internationally as the "Wizard of Finance."

He'd struck gold with an ingenious, yet deceptively simple—and, in the beginning, apparently legal—idea for multiplying money: buying International Postal Union reply coupons abroad and, taking advantage of favorable rates of foreign exchange, swapping them for their face value in postage stamps. (For example, in Spain at the time such a coupon cost the equivalent of one U.S. cent; in this country the coupon could be exchanged for 6 cents' worth of stamps—a 500 percent return. Translated into dollars, there seemed no limit to the profits.) It was the kind of get-rich-quick scheme people could relate to, poor and well-placed alike. And when Ponzi, who found he was a glib salesman, began to talk it up and demonstrate its simple virtues, it wasn't long before "investors" all over the world literally were begging him to take their money and compound it for them.

By 1920, Ponzi had headquarters in downtown Boston and branch offices throughout the Eastern states. He had bought controlling interest in the bank that handled his accounts, installed his wife and himself in a palatial country house in Lexington, was chauffeured around in a cream-colored custom limousine. Hailed joyously wherever he went, Ponzi was afforded phalanxes of city police to clear his path through the constant crowds of customers and admirers. He was like a one-man mint, and everybody seemed to share in his largess.

But then, almost as swiftly as it had made its ascent, his star plummeted. The Boston *Post*, acting on complaints by certain disenchanted investors, took a hard investigative look at Ponzi's operations and found that although his stamp-exchange plan had been legitimate and did work at the outset, it had long since become nothing more than a blind: what Ponzi had turned it into was a gigantic and very hazardous financial juggling act, taking money in with one hand and transferring it out—some to a few lucky investors, but more and more to finance his own more ambitious schemes—with the other. The *Post* discovered that Ponzi, for all his trappings and fanfare, in fact was probably insolvent. And it dug up and aired his questionable past.

Criminal investigations were begun, state and federal indictments drawn, as well as bankruptcy proceedings and lawsuits by bilked clients. Internal Revenue seized his assets (which in the end proved worthless—he was at least $3 million in debt). He was convicted on a federal charge of using the mails to defraud and in state court of larceny.

Ponzi served the two terms consecutively, almost twelve years, and was released in 1934. Then U.S. Immigration agents put him on a boat back to Italy.

Rose Gnecco Ponzi, her own world turned topsy-turvy, had stuck by her husband through all of it, including his years in prison. In 1931, with the Depression making life ever more difficult, she took a job as bookkeeper at the Cocoanut Grove. (Solomon was owner then, and Barney Welansky his attorney. After Solomon's death, Welansky kept her on.) When Ponzi was freed at last, before his deportation, Rose chose to leave him. She would carry on by herself, unburdened.

She had been a competent, trustworthy employee, doing exactly what was expected of her, and in time Welansky had made her purchasing agent for the club as well as bookkeeper. She had not remarried, and perhaps the only thing remarkable about her was that she continued to use the name Ponzi. Not that Rose cared to talk much about that dizzying phase of her life, or about her former husband.

Now the world, it seemed, was once more staring at her, wondering darkly whether it might be significant that Rose *Ponzi*, of all people, had been entrusted with the finances of the Cocoanut Grove.

But from her testimony, and her demeanor under questioning, it soon became evident that there was no significance, only odd coincidence, in the tarnished name she bore—of which she, above all, was only too painfully aware. Rose handled the books; balanced income and outgo; paid bills and the help; deposited receipts. It was elicited in the hearings that at Welansky's instruction she'd maintained a club account in her name at the Pilgrim Trust Company—a simple matter of facilitating her accounting. Beyond these routine responsibilities, she did not know, had not concerned herself with, any other aspects of the Grove's operation, professional or personal. It was a job; sometimes she'd had to work odd or long hours, as was the nature of the nightclub business (the night of the fire she'd been at the club until about 7:30 P.M.), but it was a living. Mr. Welansky had always been a decent employer. She couldn't add any more.

So nothing very illuminating or relevant to the central issue came out of the much publicized appearance of Charles Ponzi's former wife. In a way, Rose's contribution was about the same in net effect as Jimmy Welansky's: negative. But in her case, people for the most part—even those who were disappointed—believed her.

As with most such protracted hearings on grave matters, the significance or drama did not build vertically, one enlightening development upon another, but cumulatively. In addition to the city officials and Cocoanut Grove principals who got the main spotlight, many other witnesses were called, well over two hundred before it was over. They came as they were available, in no planned order of relevance, to be mined for whatever personal or pertinent abstract contribution they might add to the total picture.

There were dozens of comparatively undramatic factual witnesses, people who might not have been directly involved in the fire itself but whose knowledge of or association with the Grove could provide shades of "expert" testimony: Bodenhorn, the architect, and his builder, Rudnick (neither of whom had been present that night); contractors and workmen and painters, past and recent (the consensus of whose evidence was that inflammable decorations in the club *had* been "flameproofed" in some fashion or other at one time or another—all indications being, however, that such preventive treatment may *not* have been applied during the last several years); some former employees who suggested they'd never seen a full-fledged fire inspection of the place (their only clear recollections were of this uniformed "inspector" or that moseying in of an afternoon, taking a cursory look around, then either sitting with Mr. Welansky to chat and maybe have a drink or heading down to the kitchen to grab a quick bite). Then there were the straight all-business types, insurance adjusters and bank account managers and representatives of the city tax commission, all with specialized interests in the Grove as a commercial entity.

(Nor should it go without noting that out of all these there were some who made a point of speaking well of Barney Welansky, women who worked for him, particularly—the employees in his law office, the help in the club, the entertainers. A stocky, straight-backed man with a kind of clipped soldierly manner, Welansky was referred to by many as "the colonel"—though not to his face—but it was more an affectionate than an ironical sobriquet. For all his terse, impatient ways, they said, he was a fair man and true to his word and, deep down, they thought, caring. Employees who did the job they were being paid for never had any trouble with Barney Welansky; and if any had special problems of their own, he was one they always felt they could turn to. Above all, he showed women *respect*—could even be courtly in his brusque fashion—and *that* was not to be taken lightly in this day and age.)

Most touching or moving, of course, if not always the most reliably informative—for the intensity of experience often tends to exaggerate or blur true images—were the ones who'd been there: from some of those on the outside, bystanders who'd watched helpless or just horrified as the nightmare unfolded; to firefighters and rescuers, solemn and forever hardened; and the survivors themselves, both patrons and club employees, some of whom seemed permanently maimed in spirit.

There was one exception among these, one whose account was somehow disconcerting for quite a different reason: the eighteen-year-old college student Anne McArdle, who'd left her three companions in the Melody Lounge just before the fire broke out to go upstairs to the powder room. The way she told it, she may have been the only person in the Grove that night who saw nothing of the fire—either from inside or from without.

She and one other woman were in the powder room when somebody called from the anteroom to get out, and then they smelled the smoke. The other woman opened a casement window, which overlooked a back alley, and lifted herself out. Anne followed. The window was about nine feet from the ground, but there was a car parked just underneath and she climbed onto its roof and got down. The alley led behind the Grove out to Shawmut Street. She didn't see anybody around, so—and she related this with such equanimity, almost disinterest, that the packed hearing room, tensed for pain or pathos, began to stir uncomfortably—she calmly walked up to Park Square, stopped in a cafeteria for a cup of coffee, then caught a trolley back to her school in Brookline. As for the others she'd been with, she seemed to have no more concern for them than if they'd become separated after a casual evening at the movies. She didn't bother to look for them once outside because she hadn't been enjoying herself that much—she didn't really know any of them, and *they* all seemed to know one another, so she didn't worry about their missing her; and anyway the Cocoanut Grove was not really her cup of tea, *her* crowd always went to the better places like the Copley Plaza or the Parker House or maybe the Bradford Roof—so she figured it was as good a chance as any to break away. She didn't find out until much later, maybe the next day, how bad the whole thing had actually been. And when she realized, still later, that two of the others, including her date, had died in the fire, naturally she felt bad. But, Anne concluded, her tone a shrug, there was no way she could have known, much less done anything about it.

And then there was Stanley Tomaszewski.

By the time he was called to account, the common perception of the fire's genesis centered on "the kid who lit the match." Whatever contributory factors there might prove to be at higher levels of responsibility, for most people the bar boy was the proximate cause, the one individual who could be singled out. And many were vilifying him. Yet if a broad public attitude could be weighed with any accuracy, it seemed there were more who blamed him less harshly, who found it in themselves to feel almost as sorry *for* the hapless youngster as about the unfortunate part he appeared to have played in this tragedy. After all, no one could actually believe the kid had set the fire *intentionally*.

The hearing room buzzed, then hushed. All eyes and ears were strained toward Stanley as he was ushered in, flanked by the police detectives who had hustled him downtown from his hideaway in the Kenmore Hotel. He looked anything but the slovenly, despicable creature some were making him out to be: a tall, sturdy young man with good, strong features, straight dark hair that was barbered and combed, a neat suit and tie and a clean white shirt.

Clearly uncomfortable, somewhat apprehensive, Stanley sat before Commissioner Reilly and his panel, repeated the story he'd told to the police, and answered their questions. His voice quavered at times but his words were delivered with unwavering forthrightness, neither betraying self-doubt nor retreating to self-excusing alibis.

He could not guess how the fire had started. But it was his honest opinion that it was not the match that had done it.

When he was done, the spectators remained silent, examining this firm, direct, yet respectful boy-man. No one spoke as the panel of inquiry excused him, and he rose and was escorted out by the tight-lipped policemen.

No one who had heard him knew quite what to think.

Outside, however, a small, angry crowd jeered and cursed

as Stanley appeared—"Murderer!" "Go ahead and hide, killer!"—and was quickly maneuvered by the detectives into a waiting car. They swarmed threateningly close to the car before it accelerated away, and some lumbered partway down the street after it, shouting out their rage.

The police, taking a circuitous route to avoid being followed, returned Stanley to his lonely refuge at the Kenmore. At least now he had books and classwork, which his parents had forwarded from schoolmates, to occupy him. He knew he couldn't go home yet.

chapter 13

If there was a "star" attraction at the public hearings, it had to be Grove master of ceremonies Mickey Alpert. Not because he could shed light on the central issue, but because he was himself the most luminous of all who appeared. The Cocoanut Grove had been one of Boston's preeminent representations of glittery night life, and Mickey Alpert had, over the last three or four years, come to personify the Grove.

The interesting fact about him, either unnoticed or cheerfully overlooked by most people, was that Alpert himself had no specifically definable talent: the orchestra was billed as his, but he was not really a musician, and while he smilingly fronted the group ("One, two—*hit* it!") the actual leader was Bernie Fazioli; as MC, Alpert would harmonize now and then with tenor Billy Payne, and did an imitation of the stylish Harry Richman that was always popular, but he did not pretend to be a singer; he had a sparkling monologue but was not a gifted comic.

What he did have were three valuable qualities: first, perhaps foremost, he was a natural impresario with a true eye and ear for talent and the ability to blend it into a first-rate show. Second, he seemed to know or have access to *everybody* in show business, not just locally but throughout the great firmament of stars, and through his efforts the most dazzling lights had often shone upon the city. The regular Thursday "guest nights," when stars who happened to be appearing elsewhere in town (or even just passing through) would come late in the evening and be induced to get up and wing some stuff for free, had become one of the Grove's biggest attractions. And last but in no way least, Alpert was possessed of a natural flair, the uncommon knack of winning an audience without personally *doing* anything—call it presence, pizzazz, whatever, he had that. In Boston, then, and through much of the region, thirty-eight-year-old Mickey Alpert was a bona fide celebrity in his own right.

But despite his "star" status, since the fire even Mickey had come under an unaccustomed cloud: some questions were being raised in certain quarters, and mean little rumors passed, that he might have comported himself less than manfully in the midst of the crisis. Some said they understood he'd run downstairs as soon as the fire broke out and hidden in one of the refrigerator lockers off the kitchen until it was over. Others whispered that wherever he'd run, he'd bowled over people right and left to save himself. (Never mind that all too many others caught in the firetrap that night seemed to have acted in much the same way; from *him*, somehow a nobler performance was expected.) And what bothered many people was the newspaper photo taken of Mickey outside the Grove, dazed but evidently unhurt, wearing a woman's white fur coat over his shoulders. How had he come upon that coat? And how *had* he gotten out?

So his appearance at the hearing was eagerly anticipated. He had been among the first invited to testify, but had been so shaken by the tragedy that he'd retreated to the suburban home of his elder brother, George, a prominent attorney, to pull himself together in seclusion. Finally emerging, Mickey, normally a tall, natty, swaggering figure, seemed but a shadow of himself: stooped, dressed indifferently, with a soiled bandage around one hand and wrist; fatigue weighed heavy on his expression, his eyes were sunken in gloom. His voice was gravelly, his response to questions sometimes halting, as though he were distracted.

Those who knew him well, seeing Mickey Alpert that day, thought he looked as though he'd aged fifteen years.

Commissioner Reilly asked him gently if he wanted to talk about the fire. Alpert rubbed a hand across his eyes and murmured he'd try, as best he could remember...

They were supposed to go on the radio at 11:15. The show was late starting. There'd been a big crowd since early in the evening, and they were still coming in, and more tables were being set up on part of the dance floor—

Reilly cut in: Would he say there were more people than could be handled comfortably?

Alpert frowned and shook his head. He wouldn't say it was especially unusual. Saturday nights were always heavy. The only thing that bothered him was that the tables on the dance floor would cut down the size of the rolling stage, which came out from under the bandstand, where the girls did their dance and production numbers.

But there was nothing to be done about that, and he recalled saying to Billy Payne, "I'm bushed, let's go find a place to sit till they're ready." Hardly had they stepped down from the stand than someone beckoned to him from the Terrace section to come and meet Buck Jones. He started over—but then there was a flurry out in the Foyer and he thought he heard a shout, "Hey, a fight!" The first thing he did was look around at the band: they were under strict orders not to make a fuss if there was any trouble, to keep cool. None of them had moved; they were all looking out toward the Foyer—

What kind of "trouble"? he was asked.

Oh, every once in a while, in a place that size, with a lot of people bumping into one another, there might be some friction, a dispute over something or other—it happened, never anything serious. He just didn't want the guys up there on the stand rubbernecking, calling more attention to it....

When he turned back, he stopped dead: there was a flash of *fire* at the entrance to the room... high up, like a giant balloon floating in, all flame and smoke... and all at once people were shouting and screaming, jumping up from tables and overturning chairs, and they started coming toward him in a wave, really wild....

Alpert's voice broke, and the room was quiet.

He didn't know what to do. He remembered raising his arms and shouting at them, "Hold it! Wait a minute, now! Slow down!" But it was no use: they poured over him. He was swept along to the end of the Terrace, to the head of the stairs from the kitchen. For a moment he managed to grab hold of The Terrace's iron railing, but then he was shoved down onto the stairs... and then he remembered the back stairs, across the kitchen, and he kept on going down, with people pushing behind him, yelling, and he ran through the kitchen—

Were there others then in the kitchen?

Quite a few.

Any fire there?

No, but smoke.... At the top of the back stairs, which came out behind the stage, a bunch of people were pounding at the exit door, fighting to get it open. Smoke was rolling down on them, and there was a crackle of flames. *We're trapped!* he remembered thinking.

He saw a door to the right leading to the building where the performers' dressing rooms now were, and he rammed through it into a little unfinished room between the club and the adjacent building, and people piled in after him. Smoke and flames were already starting to fill that room too. There was a small window facing Shawmut Street. It was bolted. He smashed it with his fist and started hauling and shoving people out. Everybody was clawing at one another....

Alpert paused, his expression clouded, eyes distant. He couldn't remember getting out himself. "All I know is, I came to on the sidewalk, and John Walsh [the Boston Civil Defense director who'd been in the Grove] was slapping me across the kisser. It was cold...."

They gave him a moment to compose himself. No one had the heart to ask where or how he'd gotten the woman's fur coat so widely speculated about. Commissioner Reilly then turned to the subject of the Melody Lounge and the New Lounge. Did Mickey have any say in *their* operation? Had he been consulted, for example, about plans for the new addition?

No, Alpert said. His only concern was the main club room, and then only with respect to the show, staging and so on.

As for the locked doors and other concealed exits, Alpert shrugged tiredly. That was standard, he thought—at least since he'd been back. He'd never thought about it one way or another. He supposed it was to keep customers from skipping checks. It had never been a problem before....

Questioned about "gangsters" frequenting the Grove, or other illicit activities there, Alpert scoffed. Guys like that came in sometimes, sure—a place like the Grove got all kinds. But it wasn't any "hangout." In fact it was more a family place. Nobody could say the entertainment there wasn't 100 percent clean, *never* any smut. He, for one, wouldn't have stood for it. The only thing maybe he personally didn't care for too much was the practice of having the show girls come down sometimes and sit with certain male customers—but even that wasn't illegal, no hustling or trying to push high-priced drinks (like at some places); the girls were just there as company for good spenders. Mr. Welansky felt it was good for business. Okay, it was his joint...

About Barney Welansky, Alpert would say only that he was a smart businessman, ran a first-rate operation, had been a good employer. "He gave me pretty much a free hand on the entertainment side, and I never poked my nose into any of his other affairs. I have no complaints about Barney Welansky."

Reilly asked: Did Mickey, or any member of the Alpert family, have, as was widely believed, a financial interest in the Cocoanut Grove?

Mickey hesitated, as though having to consider his answer. Then, shaking his head, in a barely audible voice, he said: "No, sir...."

* * *

What many of the general public did not know, or had forgotten by 1942, was that Mickey Alpert had at one time been much more deeply involved in the Grove than simply acting as its resident headliner. A financial interest? No more, it was true, not for a long time. But once—

In fact, he had been one of the Grove's *founders*, and it was he who had struggled to keep it going during the lean, tough years of Prohibition. When he finally couldn't do it any longer, to his everlasting regret he'd had to bail out—for peanuts!—and then he'd left Boston. When he did return six years later, the Grove had become a highly successful enterprise. By then he, too, was a polished performer, and when he'd been given the chance to come back—as a paid employee—he'd done his share, and more, to build on that success and gain the recognition that had eluded him before. He'd actually come to think of it as *his* place again. So now, this sudden utter destruction—it was almost more than he had the strength or will to bear.

Fifteen years it had been since the Grove's inception, and even that had been bizarre.

Back then, in 1927, Mickey—born Milton—was twenty-three, out of Boston U. and incurably infected with the show business bug. He didn't yet know quite what direction to follow, but he was full of verve and wit, and he had taken up with a rising young orchestra leader named Jacques Renard (*né* Jacob Slaviski) who had already gained something of a reputation accompanying the great Eddie Cantor on some of his radio broadcasts. Mickey had joined the Renard ensemble, acting as the warm-up personality—bits of bright patter, a little song-and-dance stuff, audience participation. The orchestra played all the popular dance tunes, and the "act" had come into some demand at local clubs and ballrooms, not only around Boston but at resorts all over New England.

Early that year, while appearing at Rangeley Lakes in Maine, Mickey and Renard were approached by a well-dressed guest who was so taken by their performance and whole youthful dynamism that he made them an extraordinary proposition:

he had come into considerable wealth, he said, having recently made a "killing" in a California investment venture; and now he wanted to turn some of that toward realizing a lifelong dream—to own a lavish nightclub. He had looked around and concluded that one major city ripe for such an enterprise was Boston. Alpert and Renard were from Boston and obviously knew the territory, and he liked their style immensely. So he would finance the building of such a club if Mickey and Renard would undertake to put it all together, perform there, and run the place for him.

The two young men could hardly believe the proposal: talk about entertainers' fantasies come true! Still, they were wary; here this stranger (he identified himself as Jack Bennett, but gave no other particulars) comes to them out of the blue like a fairy godmother and offers them a deal that seems almost *too* fantastic. They told Bennett they would have to consult with their advisers. He gave them the business card of his New York attorney and urged them to accept. It was the chance of a lifetime for the three of them.

Mickey and Renard hurried back to Boston to see Mickey's lawyer brother. George Alpert, at twenty-nine the eldest of six, had assumed leadership of the family a few years before when their father was killed in an accident. A former assistant district attorney, George now specialized in civil and corporate law, in partnership with criminal lawyer John Feeney. He recognized Bennett's attorney's firm as one of New York's more prestigious. His advice was to go as far as they could with the offer *without* committing any funds of their own in order to find out how sincere this Bennett was. Always be skeptical, be prudent, he told them.

They found an abandoned building near the theater district that had been built in 1916 as a garage and later used for film storage; it was solid, of concrete and stucco, and the dimensions were right for what their fevered imaginations conceived: an exotic kind of South Seas atmosphere—smack in the cold heart of Boston! Through some friends they contacted Reuben Bodenhorn, an architect noted for his design of supper clubs and nightclubs, and asked him to look over the building and draw up a cost estimate. Then, apprehensively, they got in touch with Bennett's attorney and gave him the numbers.

Within days, they received a check signed by the attorney against Bennett's account covering all their projected start-up expenses. They were in the nightclub business—just about!

They retained Bodenhorn and left it to him to contract for builders, plumbers, electricians, decorators. Mickey and Renard themselves spent all that spring and summer working their club dates by night, by day incessantly designing and planning. They hired waiters, cooks, a chef. They wanted somebody classy out front, a polished maître d'; through contacts they found Angelo Lippi and persuaded him to sign on. With brother George's help, they secured their licenses. The grand opening was set for October.

In all this time of exhilarating preparation, not once did their benefactor/partner Jack Bennett interfere or even come around. Strange, but at the same time appreciated. His attorney promptly forwarded necessary funds upon receipt of every invoice—adding up, as the club neared completion, to about $75,000. Bennett had just one suggestion to offer via the attorney: he felt a good name for the club might be the Cocoanut Grove, after the famed Los Angeles nightspot. They liked it: perfect!

It was really unbelievable, how smoothly it all had gone!

Deep down, they'd never stopped holding their breaths, waiting for the bubble to burst. And finally, it did one late-September afternoon, when one of Renard's musicians came in with a long face and a week-old copy of the Los Angeles *Times*, forwarded to him by a friend. There on the front page, among the photos of several men accused by California authorities of perpetrating a huge stock swindle, was the face of their "Jack Bennett." Only his name was given as Jacob Berman, and he was cited as *the* key manipulator of the multimillion-dollar scheme!

A long article accompanying the photos outlined the story to date, which evidently had been unfolding since the spring:

Early in May, authorities had halted trading on the stock exchange of a "hot" issue, the Julian Petroleum Corporation—whose subscribers numbered some fifty thousand mostly small investors—because of an apparent overissue of the stock. A small panic ensued, and the Julian Corp. was placed in receivership by the U.S. district court.

Ten days after that, an L.A. County grand jury had returned a secret indictment against unnamed individuals, with bail placed at $250,000, the highest ever set there to that time. The charge: the Julian company's stock had been overissued by more than 3.6 million shares (with a par value totaling $150 million). Then, in June, indictments had been brought against fifty-five persons named in the fraud, most of them prominent California bankers and businessmen, including filmmaker Louis B. Mayer of M-G-M. (Also involved would be famed director Cecil B. De Mille, later charged with usury for having loaned the Julian Corp. $62,000 on a forty-five-day note and then collecting $12,000 in interest.) And among those fifty-five accused was one Jacob Berman—"a.k.a. Jack Bennett"—of New York, who had been a confidential aide to Julian's resigned president and who was now described in the indictment as a professional "market manipulator."

A month later, at the end of July 1927, it had been disclosed that the number-one culprit was the man Berman, who was said to have himself pocketed more than $66 million over the past year and a half, which he'd deposited in various New York banks. Still unaccounted for was another $34 million in stockholder's money.

But Berman/"Bennett" had skipped, and there had been a nationwide hunt for him since May. The New York police, contacted by the L.A. district attorney's office, had checked and staked out Berman's known addresses there—a hotel apartment on Central Park West, another flat on East 89th Street—but had uncovered no trace of him or of his wife and two children. One report had it that the Bermans had long since fled to Europe. His attorney professed not to know where they were.

Then, just in mid-September, Berman had suddenly reappeared, surrendering himself, apparently by prearrangement, to the California authorities. This was the occasion for the news coverage that had finally come to the attention of Mickey Alpert and Jacques Renard.

Where did all this leave *them?* they worried. They'd built "their" club on the embezzler's money!

George Alpert was furious—with them, for having got themselves so gulled; with himself, for having let them; and, of course, with the dupicitous "Bennett." But the two young men were so heartbroken that George rethought the situation: After all, the man was not yet *proven* a thief. Until he was convicted, his money could be assumed to be good. At the same time, it probably would not sit well in Boston if the association became public knowledge.

George and his partner, John Feeney, the criminal lawyer, contacted Berman's attorney in New York and bade him come to Boston for an urgent conference. They met two days later at the Parker House.

To the Bostonians' complaints about Berman's unconscionable deceit, the New Yorker only sniffed that irrespective of Mr. Berman's other *alleged* activities, he had fulfilled his agreement with the young men entirely in good faith: he had financed the building of a club, acceding to all their requests to the tune of some $75,000; he had neither reneged nor interfered. So why should *they* feel threatened?

The point, Alpert and Feeney argued, was precisely good faith: regardless of Mr. Berman's ultimate guilt or innocence—and it might take a long time for such a judgment to be settled—the notoriety sure to ensue could only hurt this budding—and legitimate—enterprise, possibly destroying all the hard work and expectations these two *honest* young men had invested in it. Mr. Berman could *truly* demonstrate his good intentions by simply withdrawing now, considering all the mitigating circumstances. He could write off his own investment and let Mickey and Jacques have the club, without strings, to operate as best they could. Otherwise, George Alpert added dourly, there would be no recourse—even at the expense of his brother's reputation and future—but to bring to the attention of the California authorities this disposition of funds by Mr. Berman in the Boston area, which so far seemed to have gone unsuspected.

After much huffing and stewing, the starchy New York attorney—doubtless wishing to spare his own grossly troubled client additional legal complications—proposed a compromise: "good faith" indeed being a two-way street, Mr. Berman would agree to relinquish all prior claim on the club in return for a cash settlement of $25,000. The shrewd Yankees quickly calculated and agreed.

Mickey and Renard were elated over the arrangement, but for one apparently insoluble problem: neither of them had access to $25,000. But George Alpert and John Feeney had already come to their own agreement about that: they would lend the budding entrepreneurs $10,000 apiece; if Mickey and his partner wanted it badly enough, they would manage to scrape up the additional $5,000.

They worked like dervishes for the next month, and between their earnings and personal loans from friends came up with the $5,000 and a little extra operating cash besides.

In deference to Renard's greater name recognition, the club was registered as The Renard Cocoanut Grove, Inc. (He and Mickey were listed as "stockholders." In actuality, the *owners* were the attorneys, George Alpert and John Feeney.) And by the end of October it was set to go—for better or worse.*

---

* In May 1928, following a complex seventeen-week trial (at the staggering cost to Los Angeles County of some $300,000), eleven of fifty-five defendants in the Julian Corp. stock swindle case were acquitted and forty-two other indictments were dismissed. Only two of the promoters were convicted: the former president of Julian, one S. C. Lewis, and Jacob

There had been quite a few snazzy additions to the Boston scene that boom year of 1927—restaurants, theaters, office buildings, and three new hotels, capped by the ultra-grand Ritz-Carlton. (Plans were begun, too, for a great new sports arena, to be called the Boston Garden.) But it's doubtful if any of those premières were celebrated in more gala fashion than the debut of the Cocoanut Grove on October 27.

The *crème de la crème* of café society turned out, four hundred strong, for the Grove's opening-night party, even at $10 a head and despite the strictures of Prohibition. It was quite the most glamorous spot Boston had ever seen. Just one big square room, but designer Bodenhorn had done wonders with the old garage—breaking up the space with imaginative arches and wrought-iron enclosed terraces, and decorating it with rich sconces and hangings and the opulent overhead tent effect. The crowning fillip was, of course, those spectacular, lifelike palm trees—all more evocative, really, of a lush oasis villa than the South Seas setting originally conceived.

For a nightspot, the cuisine and service were exquisite: the kitchen under the supervision of "the famous restaurateur" Fred Ruisseau, food preparation by "Louis of Paris and his French Cooks"; reservations and seating (and decorum) in the velvet hands of suave Angelo Lippi. There was no bar, of course, and no alcoholic beverages were served, but patrons were encouraged to bring their own spirits and were graciously provided with the appropriate "setups." And on stage, the star attractions: Mickey Alpert and His Entertainers, and Jacques Renard and his Victor Recording Orchestra.

The Grove unabashedly called itself a "Palace of Amuse-

Berman (who, it came out, had been in cahoots with Lewis, even having lived with him, for years prior to the great fraud). They were each sentenced to seven years at the federal prison on McNeil Island in Washington's Puget Sound. Most of the missing money was never recovered. It would be reported later that among the 50,000-odd victims of the swindle, many of whom had lost their life savings, there were a number of suicides and others reduced to living on the public dole.

ment incomparable to any in America." And quite possibly it was. Certainly it was a rage in Boston.

At the beginning, anyway.

The Grove's popularity remained high and fairly constant for well over a year. But by 1929, business had begun to fall off, and after the ruinous stock market crash in October of that year it declined sharply. Actually, the decline was attributable less to economics than to booze—that is, the *lack* of it at the Grove. The ambience was no less glamorous, nor the entertainment less lavish—the young proprietors took pains to maintain the highest quality and taste in those areas. But for many, in those frantic final days of the so-called roaring twenties, the Grove's allure was eclipsed by the newer, more daring cabarets that had since sprung up in and around Boston, places where the excitement came from flouting the law by *ordering* forbidden "hootch" instead of having to tote one's own, where the floor shows as well as the atmosphere seemed racier, somehow more stimulatingly illicit.

Alpert and Renard were aware of their problem, which inevitably led to differences between them. Renard leaned toward making liquor available; Mickey was dubious about turning their "palace" into a risky speakeasy. As it turned out, the question was academic, for George Alpert had already decided, and dictated, that his professional rectitude as well as personal propriety forbade so much as a wink at the law. And George, while never intruding in the day-to-day operation of the club, actually still held control. If they insisted on bringing in liquor, he made it plain, he and Feeney would immediately withdraw their financial and legal support. The Grove would have to float or sink as it was.

And so, it had begun to sink. By the start of 1930, they were finding it increasingly difficult even to meet the payroll. The entertainment lineup was cut back; they had to thin out the service staff. Then Angelo Lippi departed to take a better offer as maître d' at the Hotel Touraine.

Finally Jacques Renard himself walked out, taking his

whole band with him—a desertion all the more dastardly because he went no farther than to the Grove's newest rival just across Broadway, the Club Mayfair. The Grove's future, only three years after its glorious opening, looked bleak indeed. The Alpert brothers began to wonder if they could even sell it without taking a crippling loss.

They got an offer, unsolicited and unexpected, from a source that in other circumstances they might have turned down flat: Charles "King" Solomon.

They knew very well who he was, both by reputation and because he (oddly, in view of their unwavering no-liquor policy) had remained one of the club's most faithful supporters. Solomon was an acknowledged racketeer, a gambler, a bootlegger, and generally a very hard type. It was plain that Solomon had seen the Grove was floundering, for he summoned Mickey to his table one night between shows and flatly laid out a proposition: he would pay $20,000 for the joint, lock, stock, and barrel, and what's more he would keep Mickey on salary as entertainment director. Mickey all but jumped at the deal. But first he consulted George.

His brother didn't like the idea of selling out to a man like Solomon; but it was the only offer in sight. With barely masked distaste, the attorney agreed to talk with the flashy mobster. Solomon rang in his own lawyer, Barnet Welansky (whose own credentials, George found on inquiry among the Boston legal fraternity, were surprisingly reputable), and they worked out the agreement: Solomon could have the place for $30,000—the $20,000 stipulated, and $10,000 on top. Thus George Alpert and John Feeney would get out whole, and Mickey would get a small stake. All in all, it probably was for the best. But Lord knew what the Cocoanut Grove would turn into now.

King Solomon had the ideas, the wherewithal, *and* the push to restore the glitter to the Grove.

The first thing he did was to go to the Hotel Touraine

personally and get Angelo Lippi back as maître d'. Lippi demurred. Solomon said, without fear of contradiction, that he needed Lippi's class out front, he *wanted* him back, and that was how it was going to be. Also, he was making Lippi the new president and treasurer of the Cocoanut Grove, Inc. It was a proposal Lippi could not refuse. (Soon after returning, however, Lippi found that he was the club's top executive in name—for license registration purposes—only, with no say whatever as to its management or operation. Moreover, he found that the $100 a week he'd made at the Touraine had been cut back to $75. Lippi thought about walking out... but decided that would not be prudent. Nor did he chance complaining. He just stiffened his upper lip and hoped the situation would improve.)

Next, as George Alpert had anticipated, Solomon boldly set about recasting the Grove as the highest-class speakeasy in town.

And then he jazzed up the entertainment program, instituting a "star" policy unprecedented in Boston, bringing in some of the biggest and highest-paid performers. Sophie Tucker (at $3,500 a week), Helen Morgan ($2,500), and the legendary Texas Guinan and her "gang" ($5,000 for the ensemble) were among the greats who played the Cocoanut Grove over the next couple of years.

Solomon's proprietorship actually was supposed to be "silent," but it was more than an open secret, for he relished every moment of his new avocation as host. A burly, swarthy, glowering man, he reigned at his special table near the stage, a conspicuous presence in his expensive dark suits, white-on-white shirts, and bejeweled cufflinks and stickpins (alternating at times with white flannel trousers and white buck shoes even in midwinter), nursing giant Cuban cigars and acknowledging with obvious satisfaction the homage of his subjects and guests.

The Grove became once more not only *the* hot spot in Boston but, soon, a mecca for luminaries from all over America, from every walk of life—the biggest-name entertainers, star athletes, famous writers and composers, business and even government leaders.

Mickey Alpert, through all this dazzling resurgence, had continued to front the house orchestra and do his modest turn as master of ceremonies. But it hadn't taken long for him to sense that he was out of his league. He didn't fit in with Solomon's grandiose style, and he felt Solomon thought the same and that his days were numbered. Only twenty-eight, Mickey began to think it might be wise to move on, broaden his horizons, get his own act back together—and in the doing, to relieve himself of the overriding burden of insufficiency, to say nothing of the uneasiness of illegitimacy.

He had met an attractive young dancer, nineteen-year-old Kathryn Hayman (who used the name Kathryn Rand), when she was up from New York appearing at the Westminster Hotel. They'd been seeing one another and talking about maybe working up an act together that they could take on the road. Prohibition was on the way out, and legitimate clubs would be opening again across the country.

So, late in 1932, Mickey Alpert left the Cocoanut Grove (with no objection from King Solomon). It was a wrench, but it seemed the only smart thing to do. But within weeks, he would have cause to wonder if he'd made the right decision.

Early on the morning of January 24, 1933, after the Grove had closed for the night, King Solomon and an accustomed entourage went out on the town and wound up at an after-hours speak called the Cotton Club on Columbus Avenue in the rough-and-tumble South End. And there, in the men's room, the high-rolling racketeer was shot dead by a group of toughs.

It was only four weeks before the United States Congress, under the new administration of President Franklin D. Roosevelt, would overwhelmingly repeal Prohibition.

And a short time later, the late King Solomon's attorney, Barnet Welansky, took control of the Cocoanut Grove.

* * *

When Mickey left the Grove, he and Kathryn first went to New York, which was her home. He got them bookings in a few small places there. They were not strictly an "act," for Kathryn was an acrobatic dancer with training in ballet, while Mickey, with his engaging fast patter, his Harry Richman take-off, a time-step here and a bit of soft shoe there, was a sort of jack-of-all-entertainment, a social director, a *tumler*, in the Yiddish show-biz parlance. And after a while they took themselves to the road.

They played a lot of towns during the next six years: Philadelphia, Cleveland, New Orleans, New York occasionally, Atlantic City; and in time they gained appreciative followings in certain places, so that after a couple of years they could almost count on bookings according to the season—winters in Philadelphia, say, at one hotel supper club or another, summers in Atlantic City at some cabaret on the boardwalk. And they matured together, both professionally and in their personal relationship. Both assumed they would marry eventually, but not until they'd got all *this* out of their systems.

One thing Mickey was never quite able to get out of his own system was Boston. And the Grove. Every so often he would get news: how, after King Solomon was killed, the club had slid straight downhill in the hands of Barnet Welansky (Mickey only dimly remembered Welansky as the inconspicuous little lawyer who'd come around occasionally, always distant, unsmiling, barely sociable, and he'd often wondered—as had many others—how somebody like *Welansky* had got control of the place), so that by the mid-1930s it was practically bankrupt. But then, the stories went, Welansky had managed to turn it around, even without Solomon's fabulous "star" policy, and over the next few years, it was said, the Grove had made a great comeback, flourishing with just the sort of entertainment that Mickey and Renard had been putting on in the beginning, but with the advantage of open bars and booze. And Mickey sometimes daydreamed about how it would be for *him* to be back on that stage at the Grove.

Then he heard on the grapevine that Welansky was not

entirely satisfied with his shows, might, in fact, be looking for a new entertainment director/master of ceremonies. And so, early in 1939, during a break in their schedule after the holidays, Mickey decided to pay a long-overdue visit to his old hometown.

He'd hardly stepped off the train at South Station and breathed the fishy salt air off the nearby wharves and the harbor, when he realized how much he'd really missed Boston.

After a few days of reunions with brothers and sisters and old acquaintances, Mickey went to the Grove. The club looked much as it had—except that they'd expanded the Foyer and added a cocktail lounge downstairs—and it *was* doing good business, as crowded and gay (well, almost) as he remembered from the early days. Some of the same employees were still there, too, most prominently Angelo Lippi, appearing more imperious and fussy than ever, and Mickey got a grand welcome. Lippi took him inside to say hello to Barney Welansky.

Welansky asked what he'd been doing, and Mickey told him and asked how his entertainment was shaping up. Welansky indicated he'd been having some problems there, and invited Mickey to sit with him and watch the next show.

Afterward, Welansky asked him what he'd thought. And Mickey told him straight out: the orchestra was not half bad but it needed a little more zing; the featured acts, the production numbers, fast and loud but pedestrian, could use some imaginative sparkle; the new Irish tenor was solid; the MC—pleasant enough, but forgettable; no pizzazz! What Welansky needed, he said, was someone who could both put together a *first-rate* revue, outshining any other in town, and get up there to make it sizzle.

Welansky digested all this without comment. But later, as Mickey was leaving, he couldn't help noticing how many of the patrons recognized and went out of their way to greet the smiling, confident figure making his way across the room. And how Mickey himself seemed to expand, clasping hands, clapping people on the shoulder, leaving everyone beaming.

And a few days later, before Mickey returned to New York, Welansky asked him if he would like to take over as entertainment director at the Grove.

Mickey did not hesitate: only if he had *complete* charge of the shows.

Agreed, said Welansky.

Then it was a deal, Mickey said. He would need a couple of months to fulfill some prior commitments, then he would come back and make the Grove the palace it was meant to be!

So in the spring of 1939, Mickey Alpert came home—to Boston and "his" Cocoanut Grove. He was thirty-four.

The only discordant note was that Kathryn did not come with him. He promised she would be a permanent featured dancer in his new shows; but she wasn't ready yet to settle in one spot, especially not in Boston. She was only twenty-six and still had a lot of dancing left in her legs, and she wanted to stay on the move. Kathryn thoroughly understood his choice, even, knowing the depth of his attachment, encouraged it. It wasn't as if they were calling it quits; Boston and New York were only two hundred miles apart; they would visit each other regularly.

Over the next three and a half years, the reunion of Mickey Alpert with the Cocoanut Grove grew into a veritable love feast. He refashioned the shows and put together an orchestra of his own choosing, and irresistibly the place truly became *his*, in spirit and flavor and in the public conception (*Mickey Alpert Presents*——!). He had his band on the radio every week. He took the troupe out to military camp shows and bond rallies downtown. The Cocoanut Grove *was* entertainment, and *he* was the Cocoanut Grove. And he reveled in it all.

Barney Welansky—with no ambition to be a showman or even a "host," but knowing all about the performance of a balance sheet—reveled in Mickey's acclaim as well. Having acquired the Grove for a song from King Solomon's widow in

1933, he discovered, only a couple of years later, that he couldn't give the place away (in 1935–36, anybody could have had it outright for $45,000). And now, Welansky was turning down offers of three and four times that amount. By 1942, the Grove was *netting* $5,000 a week—a quarter of a million a year! And with the war on, business could only get better. That's why Welansky had added the New Lounge at the Broadway corner (just across from the rival Mayfair), why he had further plans for enlarging the show room by building over the small parking lot, and why he envisioned buying up more property along Piedmont and Shawmut streets, so that one day the Grove might take up the entire square block— a solid block of gold!

But then, that terrible Saturday night, in one flash of devastation, it all had dissolved—all the gold, all the dreams, of both Barney Welansky and Mickey Alpert.

Kathryn, in New York, had heard about it on the radio early Sunday morning. Frantic with worry, she tried Mickey's number in Boston over and over, but there was no answer. Then he called her. But it was not the Mickey she knew: he sounded tortured, bewildered, feeble. He was at George's house in Newton. He—could she come?

Taking the next train, Kathryn found him in a pitiable state: ashen, unkempt, wounded, frightened. And being overprotected: the authorities wanted him to appear at public hearings; George intended to spare him that. This dumbfounded Kathryn, then aroused her. Why should Mickey be kept incommunicado? Did he know "secrets" about the Grove that could possibly discomfit George and his family *now*? To most, Mickey *was* the Cocoanut Grove. If he hid, people would only assume there *was* some reason for it. He *had* to appear!

So she had persuaded Mickey to pull himself together and go to Boston to the fire commissioner's public inquiry. He had managed to tell what he knew. And when he was finished he'd shuffled off, his once proud frame bowed, Kathryn leading him as if he were a sleepwalker.

In the days that followed, it became apparent that some, maybe many, among the public anyway, still questioned his account—of how *he* had escaped; of his disavowal of any proprietary interest in the Grove (and therefore of a share of the blame). And that was the toughest of all to take.

Mickey could no longer stay in Boston after that. All that was left for him there was the ashes of destroyed hopes. With Kathryn he would go to New York and try to renew himself.

Sadly, the image he would leave behind—overlaying all the glitter and applause—was the picture of a dazed man in a disheveled tuxedo, supported by begrimed rescuers outside the ruins of his own pleasure palace, a woman's white fur coat incongruously draped across his shoulders.

chapter 14

While all the public melodrama was taking place center stage, being reported on the front pages and over the radio, out of general view more personal and even more intense dramas continued to be played out privately in hospital wards and operating rooms.

While many victims of the Grove fire responded well to treatment and were discharged within days or weeks, others—several score, overall—would remain hospitalized much longer, some well into 1943. A few of these would not ultimately survive.

As results were tabulated, toward the end of December there grew at Mass General (where no patients were lost after the first seventy-two hours of crisis) a particular sense of pride— a quiet but discernible satisfaction in perceived achievement. Of the seven who had succumbed soon after admission, all had suffered irreparable pulmonary and other internal injuries; two of those had also sustained extensive burns, but in neither case had burn damage been the direct cause of death. All other severe burn patients were recovering nicely, and some had already been sent home.

Thus while Mass General's mortality rate in this emergency was about 18 percent (seven of thirty-nine), its percentage of burn deaths was zero. By comparison, at Boston City Hospital the losses were closer to 30 percent (thirty-six fatalities by the end of December, and several more likely, out of the original 132 admissions), with a number of these wholly or partially attributable to "full thickness" or third-degree burn damage.

But comparative statistics alone were hardly what prompted gratification at Mass General—true medical professionals could not be cheered by patient losses anywhere, whatever the ratio. It was, rather, engendered by the knowledge that the innovative burn therapy applied there for the first time in a general crisis had proved to be a virtually unqualified success. And this judgment had been formally endorsed by the National Research Council, which had had observers closely monitoring the effectiveness of the emergency burn treatment at both the local hospitals funded for such research. The council now had pronounced the "soft" technique devised by Dr. Oliver Cope and his colleagues to be demonstrably superior to the old methods still practiced at Boston City and elsewhere, and worthy of immediate promulgation. The so-called Cope method, then—with its relative deemphasis of *surface* burn treatment in favor of all-out concentration on *internal* treatment to avert infection and shock—was ordered to be written up in detail for early publication in medical journals, whereupon it would receive the council's imprimatur and official recommendation for immediate adoption by all the nation's military services.

But, of course, elation over this historic achievement did little to ease the pain of those still trying to overcome the ravages of their experience—at Mass General as anywhere else.

One of the badly hurt survivors there was Martin Sheridan, the free-lance journalist who had been Buck Jones's press representative in Boston. Sheridan had been brought in semiconscious. His hands were seriously burned, with lesser burns all over his head and face, including the corneas of his eyes, and he was having some difficulty breathing, indicating inhalation burns as well. He recalled being asked if he knew his blood type and, having recently given a pint to the Red

Cross, replying that he should have a card with the information on it in his billfold. (This, he would learn later, probably had saved him about a half hour's precious treatment time. Too many people did not know their blood type and had to be tested before critical transfusions could be administered.) Then Sheridan had been sedated... and that was the last he'd been aware of for three days.

When he awoke finally, Sheridan was struck with panic: both his hands were encased in thick, moist bandaging; his entire head was bandaged; he could not see! Oh God, he despaired, a writer's only tools—his eyes and his hands!

But attendants quickly reassured him: his eyes should be all right, they would be uncovered soon; his hands would require surgery, but in time they, too, should again be functional.

When he'd composed himself, Sheridan thought of his wife, Connie. Heart skipping, he lay there numb, scarcely able to breathe, trying to bring himself to ask... but couldn't. They would tell him, surely—that she had made it out, was all right. But nobody said anything, he couldn't judge for how long, and finally he croaked out the question. A gentle voice said she was dead.

He'd known it, of course. He wept silently in his darkness, tears stinging his burnt eyes beneath the bandages; and later he felt totally drained.

Buck Jones's old friend, producer, and personal manager, Scott Dunlap, had been delivered to City Hospital in critical condition, suffering mostly from smoke and heat inhalation. It was several days before he regained consciousness. His survival hung in the balance for a period of weeks, as the doctors battled to control the accumulation of fluids in his lungs. But the crisis was eventually overcome; he would pull through.

As he slowly regained strength, thinking of the long, sad journey home to California that lay ahead of him, Dunlap asked for an accounting of his effects, whatever may have been salvaged from the fire. He was most concerned about his wallet, which had contained both valuable personal papers and a considerable amount of cash.

His wallet—suddenly he remembered:

Dunlap's first sensation, after having passed out on the Grove's Terrace from the overwhelming fumes and smoke, had been of the cold, wet sidewalk on which he lay. He could hear many feet tramping, running all around him, and loud, harsh voices. He was barely able to move; it was as though he had been anesthetized. One of the voices came closer, and Dunlap could hear: "That one's gone...he's dead...she's breathing...forget him..." and then the voice seemed to be moving past him. With great difficulty he turned his head enough to see that he lay amid what seemed a solid mass of prone and crumpled bodies. Stabbed with intense fright, he choked out: "Hey, please, I'm alive!"

He sensed somebody come and bend over him and heard a male voice say, "Here's one."

Dunlap felt himself being turned over on his back. He heard himself plead, pitifully: "I'll give you three hundred dollars to get me to a hospital right away!" He could not even make out the face of the man, just a begrimed shadow hovering over him.

The voice rasped: "Where's the three hundred?"

"In my wallet," moaned Dunlap. "Please, just take it!" He felt hands pawing inside his jacket...then, after a moment, he was being lifted by strong arms. And then he blacked out again.

He'd had over $800 in his wallet, Dunlap remembered in his hospital bed. Well, he thought fatalistically—and with some shame at the memory of his actually having offered somebody money, at a time like that, to save *him* before others—he couldn't really complain about losing it all. A nurse brought him his wallet, singed but intact. And when he opened it—to his utter amazement he found over $500 still there!

The guy, whoever he was, had actually taken the $300, but *only* $300! Human nature. Would it ever cease to be unaccountable?

Also at City Hospital were the two Grove musicians injured badly enough to be hospitalized, bassist Jack Lesberg and saxophonist Al Willet. They, with violinist Bernie Fazioli, had fallen to the toxic fumes by the locked Shawmut Street door behind the stage, and all three had been dragged out by firemen. Why Lesberg and Willet lived while Fazioli did not was inexplicable.

Neither had serious external burns; the doctors guessed that other victims had fallen on top of them, protecting their bodies. But each had suffered deep lung damage—which, especially for Willet, a woodwind player, could have discouraging consequences.

All Lesberg remembered of his rescue was coming to out on the sidewalk and through his haze seeing a priest about to give him last rites. "No, no, not *me*!" he'd croaked, before drifting off again.

Willet had awakened in the hospital, in bed, needles and tubes seemingly stuck into every part of his body. Dopily he'd tried to focus: a large area, filled with beds; everything white—except the figures in the other beds, all of whom appeared black. Strange, he'd thought, that everybody else here was Negro. (What he'd seen, he would realize later, were the blackened bodies of critical burn victims.) Then he fell unconscious once more and did not awaken again for three days.

Neither Lesberg nor Willet, placed in different wards, would know for weeks that the other had survived. At first, they were both given heavy dosages of morphine, but when the injections were diminished after about a week, Lesberg for one experienced a terrible anxiety: he felt he *had* to have it, and he began to lie about how bad his pain was. The doctors, knowing what was happening, cut him off altogether, and he had to pull out of it suffering a dual agony.

It was a week before Willet understood how awful the

tragedy had been. A visitor (his sister had come down every day from Newburyport, thirty miles away, and his sister-in-law, and Pepper) had left behind a several-days-old copy of the Boston *American*, and, finally aware enough to concentrate on the blur of type, he'd picked it up and been confronted by an endless list, page after page with names and little photos, of Cocoanut Grove casualties. (No one had once mentioned it to him!) In stunned disbelief, he found himself scrabbling through the pages, fearfully searching for one particular name.

It was there, among the H's: William Harlow—dead, Southern Mortuary. Willet cried out in remorse.

Billy Harlow was a delightful young fellow, in his early twenties, from a family of wealth and breeding, who'd come often to a private club on Beacon Hill when Willet played there. Al had befriended the likable young man, who loved music and wanted to be a singer. He was very good-looking and had a pretty fair voice, and after Willet started at the Grove he'd arranged an audition for Billy. Mickey Alpert had liked him (figuring, no doubt, to attract more of the society crowd), but Billy had never sung at the Grove because a few months into the war he'd joined the Merchant Marine. They'd kept in touch, and recently Billy had written from his base at Portland, Maine, that (a) he'd gotten engaged, and (b) he thought he would be shipping out soon—who could say for how long? He and his fiancée were coming to Boston for the Thanksgiving weekend, and he wanted them to meet. Al had replied that if they came to the Grove the evening would be on him.

They had come that Saturday for dinner and the first show, and later sent word to Al backstage. But he had gotten embroiled in a hot card game during the break, and sent a message back saying he was tied up but to stick around and they would all go out after the last show. Billy had sent his waiter again to say they really wanted to get away early, couldn't Al come out just for a few minutes before the next show? But Al had kept putting it off—the cards were falling for him, he was on a streak; he really did want to see the kid and meet his young lady, but hell, they could stay awhile longer. And before he knew it the time had come to get ready for the 10:15 performance, and there'd been no chance to find them out front; he hadn't even been able to spot them from the bandstand in all that mob. He'd only hoped they hadn't left yet. They would have a great time together later. But then, moments later, all hell had broken loose, and—

And now there was Billy's name. *Dead, Southern Mortuary.* He couldn't even remember the girl's name. Were they both—? Because *he* wouldn't lay down his cards. *I bet one life and raise one more*—! Al began to cry and couldn't stop. (For a long time afterward he would dream of Billy Harlow, and a young woman he'd never known, and would weep a thousand times more.)

One of the more serious burn patients at City Hospital was, ironically, a ward of Mass General's Oliver Cope.

He was a youthful Jewish refugee from Germany named Ferdie Bruch, whom Dr. and Mrs. Cope had helped to flee the threatening Nazi regime in the mid-1930s, when he was sixteen, bringing him to live with them in Cambridge. He had gone on to Harvard, been graduated the previous June, and had plans to become an architect. But with America in the war he had volunteered his services to the U.S. Army as an interpreter, and had been called to leave the following week. That Saturday, then, Ferdie and his steady girlfriend had gone to the Cocoanut Grove for a farewell night together.

When the fire broke out, he threw his suit jacket over the girl's head and shoulders and somehow got the two of them outside—before he collapsed and they were separated. She was unhurt. But at first Ferdie was listed among the "missing." It was only after a frantic eighteen hours of searching that the girl, with Mrs. Cope, had located him at City Hospital.

Ferdie's burns were significant. The best he could hope

for, the doctors said, was the loss of both his ears and permanently deformed features.*

Another critical, and especially heart-rending, case at City was the Grove's young pianist/singer, Dorothy Myles. She had been burned over almost half her body, and most severely on her face and hands.

Dorothy hadn't been at the Grove long but had quickly become a kind of pet to all who worked there. She wasn't yet eighteen (which was against the city statutes), and nobody knew for sure how one so young had gotten the job—one story was that her parish priest, for whom she'd sung in the choir, had personally interceded with Barney Welansky—but none of them cared. Dottie, as all called her, was lovely and sweet, with soft auburn hair and a cameo complexion, fresh as a blossom yet with a spicy pertness that made her youthful innocence the more appealing. And she had talent, her voice clear and lilting, her piano style, like herself, bright and soft and unaffected. Dottie was going places, everyone had been sure of that.

Saturday night, she'd finished her set in the New Lounge, turned the piano over to the older pro, Maxine Coleman, and gone to the powder room to resume studying her algebra (she hadn't even completed high school yet!)—when the firestorm erupted. Rushing out, she was caught up in the wild stampede and overrun. She'd lain among a pile of corpses until firemen, clearing the dead bodies, saw a movement of her arm. Her nose was broken, and the rest of her face was blackened pulp.

At first her general condition was bad enough that she was given only a fifty-fifty chance to live. Yet she held on, even surviving almost fatal shock. City's best medical prognosis was that there was practically no hope of her ever being able to perform in public again. It might be possible, with

---

* Ferdie Bruch *would* overcome these disfigurements and later go on—under a different name—to achieve distinction as an architect.

many operations, to reconstruct her hands so that she would play a piano, after a fashion; and perhaps she could one day sing, for her lungs and larynx, while scorched, had escaped permanent injury. But would she ever *want* to perform again? Her face was terribly damaged, and it would require almost endless plastic surgery, months or years of excruciating grafting, to restore even a semblance of normalcy.

And, aside from the potentially crippling psychological damage, Dottie had another physical problem: her valiant fight to hang on to life had weakened her heart, which would require even longer, discouraging rehabilitation. Would she ever regain the stamina *or* the will to overcome her personal tragedy? Could anyone?

Another of the most serious burn cases was Bill Shea, a 385-pound Grove bartender, also from the New Lounge, who'd sustained full-thickness burns of his back, shoulders and all extremities. Nine pints of blood had been pumped into him the first night alone. Stretched out on his stomach beneath a bed attachment called a Bradford frame—which kept even a sheet from touching his raw back—totally immobile, fed intravenously, Shea required three special nurses to tend him, and when he regained consciousness, he needed massive injections of morphine for the unbearable pain.

But life continued to ooze from his great bulk. He was given no chance.

In worse condition even than Shea, with absolutely no chance to live, was the heroic young Coast Guardsman, Clifford Johnson—burned over nearly three-quarters of his body area.

From the moment he'd been brought in, it was obvious to the City staff that this gruesome wreckage of a man could not survive more than a few more minutes. Much of his lanky frame was charred as black and rough as leather, in some places the flesh and deeper tissues destroyed down to bone or within a thin layer of internal organs, so that it was almost impossible to find a spot on an arm or leg to take his blood pressure; only his boyish face—except for one burn on the chin—and his hairless chest had somehow been spared the consuming flames. What was amazing was that he'd made it to the hospital still alive: his heart was a faint trip-hammer, he was barely breathing. Nobody could survive such massive trauma; nobody ever had.

They did what they could for him—injected morphine, sprayed exposed tissues thickly with violet triple dyes to try to hold in the exuding fluids, infused plasma—but there were too many others less badly injured, who might yet be saved, to allow undue attention to one so surely doomed. Johnson was consigned to the care of a newly graduated medical student.

The minutes lengthened into hours... a day... days, and Johnson impossibly clung to life. He took seventeen pints of blood plasma in the first twenty-four hours, four liters of saline solution, dosages of sulfadiazine against spreading infection. And—though his temperature fell lower and lower, his skin turned clammy, his respiration and pulse were still barely measurable—he continued to hold on, if only to a gossamer thread over a bottomless chasm. Yet he remained in deep coma, and those who watched his remarkable continuance had to believe that each day must be his last.

Caring for Johnson had seemed at first a thankless task to Philip Butler, the Tufts medical graduate working at the hospital as a clinical clerk while waiting to begin his internship. But Butler stayed doggedly at the sailor's bedside for seventy-two straight hours, constantly dabbing the sera escaping the boy's ruptured body, spraying his wounds with the dyes, infusing sustaining fluids—and to his own growing surprise, and with faintly budding hope, keeping his patient alive.

On the fourth day, with the hospital workload having stabilized, the extraordinary case drew more expert attention. Johnson was examined by Dr. Charles Lund, in charge of the burn-treatment research program at City Hospital, Dr. Robert Aldrich, late of Johns Hopkins and now City's resident burn expert, and surgeon Newton Browder. From all they could make out, there still was no logical medical explanation why this man was not long dead. But he was not—and that was a challenge too compelling to ignore.

Johnson thus was given over to the experienced hands of Dr. Browder. He could not have asked for better. Newton Browder, forty-nine, an Indianan who'd come east to Harvard Medical School from Butler University, had been at City Hospital for twenty-two years. He was not only a first-rate surgeon but a teaching doctor, and more, a remarkably caring one. He maintained a private practice out of a small office on Commonwealth Avenue, but devoted most of his time and energy to his hospital patients, who were generally needier. (He'd stopped attending Sunday-morning church services with his wife because he felt he could do more good at the hospital at that hour.) And now Browder was about to devote himself wholly to this terribly ill young man who had no chance.

Browder first had Johnson isolated from other patients, placing him in a four-bed room from which the other beds were removed. He asked young Butler, whose attention and dedication had impressed him, to stay with Johnson whenever and as long as he could. And he arranged with the Red Cross for teams of special nurses to tend the patient around the clock. Everything that could possibly be done for this boy would be—as long as it was necessary.

From that point, the care of Clifford Johnson became a ceaseless, exhausting round of blood transfusions, protein infusions, pain-killing injections; of repeatedly cleaning his wounds, which festered again and again with infection (fearful of applying sulfa drugs for the likely destructive effect on the patient's kidneys, every day Dr. Browder himself painstakingly sterilized each lesion by hand); then endlessly spraying and respraying them with the tannic acid solutions and triple dyes—the only tested method of treatment employed at City; changing the soiled bedding every eight-hour shift; checking Johnson's vital signs *hourly*—all the while in the unspoken certainty that surely he would die at any moment. (And praying silently as well that he not awaken before he died, for his pain would be unthinkable.)

But the days became *weeks*, and Johnson, though not awakening, did not die. His gaunt form, lying face down, almost motionless but for passing spasms, shrank a little more each day—literally melting away from the continual loss of more vital protein than they could pour back into him. It was being flushed out of him both in the natural urinary process and in the relentless seepage from his massive wounds, which no amount of "tanning" seemed to stem. By the end of a month he showed a protein deficit of 2,000 grams, which in terms of muscle tissue represented a loss of some 25 pounds. They had not been able to weigh him because of his fragility, but Johnson's normal weight had been estimated at about 165; now, wasting, he was down to 140 or so, stretched ever more thinly over his six-foot frame.

Then they almost lost him to dreaded hemoglobinuria—"blood poisoning." Because so many blood cells destroyed by the fire demanded ejection from the body, healthy oxygen-bearing red corpuscles were being swept with this flood of debris through the kidneys and out in the urine. Fourteen other Grove victims at City already had died of such toxemia—none in so bad shape as Johnson. All they could do for him was to keep pumping new blood and liquid protein through the tube into his stomach and hope. But there was no reasonable hope; on top of all else, this *had* to be the end.

Yet, amazingly, his body, as feeble as it was, kept fighting back. He overcame the hemoglobinuria. But the battle surely must have dissipated what meager reserve of strength he might have had left. Once back to "normal," the life still drained out of him; he had shrunk to 125 pounds, just a porous sheet of skin over bones. Soon there would be nothing left of him to support life.

They'd learned more about the youngster's background by then. Just twenty-one, from a tiny rural community called Sumner in north-central Missouri, he'd been in the Coast

Guard only since the previous May and was stationed at Nahant, just north of Boston. The Red Cross had been in touch with his family, and as he continued to linger, arrangements were made to bring someone from Sumner to Boston to be with him when he finally died.

Clifford's mother and one of his sisters—he was one of six sisters and six brothers in a family that eked out a bare living off the land—made the long journey by train. They were put up in rooms near the hospital by the Red Cross, which also paid for their food, and every day for two weeks they came to keep vigil. Sitting quietly off in a corner of his large room, gowned and masked, watching forlornly as the doctors and nurses scurried busily about his almost lifeless body, they prayed.

Finally, however, they could stay no longer, and sorrowfully the two women returned to Missouri. But they left him still alive.

chapter 15

Maître d'hôtel Angelo Lippi, fifty-six, bedridden almost continuously since October—first with a severe lower back strain resulting from an earlier auto accident, from which had developed arthritis of the spine, and then with a bad case of gout in his right foot—still was too incapacitated to attend the inquest. He did, however, give the investigators a lengthy deposition, and later repeated his comments to reporters. Following are excerpts as quoted in the Boston press:

"I found it all right working for Barney [Welansky]. He told me I was to receive guests, see that they were made comfortable, see that they left happy, and see that they came back again...

"I was the boss of the dining room—in charge of all the help and of the kitchen. But I was never consulted regarding expansion of the club. If I had been, I would have objected to it. I would never have permitted overcrowding.... I sent away many people to avoid overcrowded conditions. I would put up a rope and just tell people there was no more room....

"If I had been there in charge that night, I never would have left the place. I would have died helping people to safety. I'm sure I would have been able to calm down a good many of the people. I certainly would have directed many of them to safety by the back way, thus avoiding the jam at the revolving door....

"In all the time I was there, I never saw any fire or building inspectors in the place. I didn't know anything about fire exits, and no one ever told me [about safety requirements]. The police would come in occasionally, ask me how things were going, pass the time of day, and then go out...."

All through the month following the fire, the joint investigation by the state attorney general's department and the county district attorney's office had raced—at times side by side, sometimes a step or two ahead of or behind the public inquest—to collect sufficient evidence to present to the sitting grand jury before its term expired at year's end. They just made it. On Thursday, December 31, the Suffolk County grand jury returned criminal indictments against ten persons named as sharing proximate and ultimate responsibility for the disaster.

Some of those charged were obvious targets. Others caused raised eyebrows. A few were totally unexpected:

—Manslaughter and willful conspiracy to violate building laws: *Barnet Welansky*, principal owner; *James Welansky*, proxy for the owner at the time of the event.

—Manslaughter: *Jacob Goldfine*, wine steward for the establishment, who was, in the absence of superiors (B. Welansky and maître d' Lippi), nominally "acting manager" that night—a surprise because his name had hardly come up to this point.

—Conspiracy to violate building laws: *Reuben Bodenhorn*, architect/designer of the club; *Samuel Rudnick*, contractor/builder of the New Lounge; *David Gilbert*, Rudnick's building foreman (another not previously given much prominence).

And then came the shockers.

—Failure to enforce the building laws, willful neglect of duty: *James Mooney*, Boston building commissioner.

—Failure to report violations of the building laws, willful neglect of duty: *Theodore Eldracher*, inspector, Fire Prevention Bureau.

—Accessory after the fact of manslaughter, willful neglect of duty: *Frank Linney*, lieutenant, inspector for the Boston Fire Department.

—Willful and corrupt neglect of duty: *Joseph Buccigross*, captain, night commander Division 4, Boston Police Department.

The twenty-man panel before retiring also issued a summary statement with the indictments. In it they presented sharp conclusions about certain conditions found to exist in the city administration, which they urged be corrected as speedily as possible:

1. We have found, among members of various departments charged with the protection of public safety, laxity, incompetence, failure to fulfill prescribed duties effectively and also lack of complete knowledge of duties.
2. We have found shifting of responsibility and a tendency by various officials in different important departments to rely too much on their subordinates without exercising sufficient and proper check on such subordinates (as well as) to attempt to shift responsibility to some other department...
3. We have found no complete coordination between Building Department, Fire Department, Police Department, and Licensing Board with respect to various types of inspection intended to insure public safety in addition to protecting the public health, morals, etc.

At their arraignments, each of the accused—including Barney Welansky, out of bed now though still infirm—pleaded not guilty. (Not appearing in court with the others was Police Captain Buccigross, who called in ill that day. He was arraigned separately.) The three major defendants, the two Welanskys and Goldfine, were each required to post a $10,000 bond; bail was set at $5,000 apiece for Fire Lieutenant Linney and the contractor Rudnick, at $2,500 for the foreman Gilbert, the architect Bodenhorn, Building Commissioner Mooney and Captain Buccigross, and at $1,000 for fire inspector Eldracher.

The first trial date, for Barney and Jimmy Welansky and the wine steward Jacob Goldfine, was set for March 15, 1943.

It was not until a month after the Fire Department hearings ended in January 1943—more than two months since he'd been secreted in protective custody at the Hotel Kenmore—that the Boston police felt it was safe for Stanley Tomaszewski to return home and to classes at Roxbury Memorial High.

The fires of hatred had died down by then, and he was welcomed back by his schoolmates with little comment. Occasionally some "smart-mouth" would needle him, but Stanley always held himself in check, not rising to the bait either defensively or in anger (for in his own mind there was not the slightest doubt that *he* had not caused the fire), and he was big enough physically to discourage most of his peers from pushing him *too* far.

But it was a state he would have to learn to live with for a long time—wariness of challenge that might come at any time from any quarter, and preparedness for a patient, measured response.

That May 1943, Stanley was graduated with honors from Roxbury Memorial and, just seventeen, entered Boston College on full academic scholarship as a business major in the accelerated wartime program. A year later, upon turning eighteen, he would enlist in the Army Air Corps and leave the haunting memories of Boston behind. But he would find that the specter did not die easily, and even when he returned, years afterward, it would rise up to torment him.

As January 1943 drew to a close, nine of the original thirty-two Grove survivors remained at Massachusetts General Hospital. Several of these were by then almost fit for discharge, and one was journalist Martin Sheridan. He had had a more difficult time than doctors had anticipated—directly traceable, they felt, to the melancholy induced by the loss of his wife.

Physically, Sheridan had progressed quite well at first: his eyes had, in fact, sustained no lasting damage; grafts had been successfully applied to his hands three weeks after the fire and again two weeks later. It was his psychological recovery that had been slower and even erratic.

In the beginning, after Sheridan had learned about Connie, he became extremely restless, jumpy, disoriented. But then, overnight, he turned lethargic, less and less inclined to do anything—disposed only to lie or sit motionless, staring into space, disdaining to communicate with others.

Early in January he was permitted to get out of his bed for the first time and admonished to walk about as much as possible to restore circulation and strength to his legs. But he would not walk, persisting in sitting in a chair whenever he was left alone.

Within a week he had pain in his left leg; it was tender and beginning to swell rapidly. The doctors found he had developed thrombophlebitis, and they quickly operated to remove a large blood clot that had been moving up through the thigh and, had it been allowed to progress much farther, could have become a lethal time bomb in his body.

That close call seemed to jolt Sheridan back to reality. He began to apply himself toward getting well; and on January 25—his fifty-eighth day—he was told he could go home.

Sheridan would have another two months of outpatient therapy to get his hands fully workable. Friends had to bring him clothing from home to wear on leaving the hospital, for all that was left of his personal effects from the night of November 28 were his wallet containing six dollars, a fountain pen, a ring that had been filed off his finger, a battered pair of shoes, one sock, and his suspenders. At dinner, someone had to break and butter his rolls or bread, cut his food, all but feed him like an infant. Sleep was difficult for a long time, and often in the middle of the night he needed someone to talk to. Returning to a set routine was awkward and sometimes discouraging. But gradually he would manage it, getting over the hardest part, Connie's no longer being there, and by May 1943—almost six months after the fire—Marty Sheridan at last felt that he could try to pick up his life where it had been cut off.

One of the first affirmative things he did then was to "return" another pint of blood to the Red Cross.

By February 1943, with the official Cocoanut Grove death toll at 488, three survivors remained critical, all at City Hospital; none were expected to live.

But one was Clifford Johnson—and no one at City had expected *him* to live beyond the first hour. Yet here he was, going into a third month of confounding all medical precedent and modern expertise. A certifiable phenomenon—literally one for the books.

The amazing youngster had even regained consciousness!

By the turn of the year, Johnson had become an obsession of the entire incredulous hospital staff. And shortly he'd begun to attract the attention of the general public as well. The media had picked up on his story and set about chronicling his survival day by day, as if tracking some epic human adventure—an around-the-world solo flight, a new assault on unconquerable Everest. People all over were *rooting* for the kid: the bets-off underdog taking an impossible shot at the title.

The doctors understood the fascination, but they did not believe in miracles. In the end, this would be but one more tragic story to come out of the Cocoanut Grove.

To that point, after all, aside from the extraordinary fact that he *was* still alive, Johnson's only "progress" had been that his hair had resumed growing—now he had a wispy mustache and scraggly beard, and his head had become shaggy—and, while still deep in coma, he'd begun to respond to the pain roiling within him, writhing and groaning more often in unconscious discomfort.

They had taken him off morphine after the first week or

so—both because of that drug's treacherous addictive power and for fear of paralyzing his already feeble respiratory system—and had been using the milder codeine ever since. But codeine itself was a powerful narcotic which, if taken regularly and in the quantities Johnson required, could also result in addiction. It was a difficult decision for the doctors: if they withheld *all* opiates, Johnson's body might react so violently to the agony, even in his unconscious state, that he would rocket into fatal shock. So out of compassion they'd chosen to risk continuing him on heavy dosages of codeine. It was only a matter of time, anyway, they'd assured themselves again.

But then, one day in mid-January, he suddenly awoke. One moment, Johnson had been lying there, on his stomach, arms spread-eagled, as motionless and insensible as he'd been since November... and the next, his eyes had fluttered open.

He'd blinked sideways, unable to move his head, and after a minute focused on the white silhouette of the astonished nurse hovering close alongside him. And he'd wet his lips and forced a guttural sound from his throat—unintelligible, but she could swear he was trying to say "Hi"—and even flexed the corner of his mouth in what could have been an attempt to smile!

Just unbelievable.

Though at first only partly rational because he was emerging from mind-numbing shock, Johnson became more aware each day. Yet, paradoxically, his general condition did not improve; indeed, it continued to deteriorate. They began to feed him milk protein orally to supplement that being pumped into him by tube, but he could not ingest enough of it to make a difference: his body was still discharging more protein than he was taking in. The continual tanning processes could not keep up with his dripping wounds; just when one area of discharge would appear to have been sealed, leakage would seep through the artifical crust at some other place on his body. He kept losing more vital fluids in urination and periodic diarrhea and vomiting. His weight dropped to 110 pounds; his blood was turning thinner; he ran a continuous high fever. And, of course, as his senses returned—which itself was dumbfounding—he was in increasing agony and required more and more codeine.

And so he'd lingered into February—wasting away, tortured, helpless, yet lucid. He told Philip Butler he wished he could die. Butler would sit by him for hours at a time, talking, trying to give him hope. And Johnson would lie there, his face screwed up, grunting an impatient response or groaning as yet another wave of pain surged through him.

A couple of weeks into February, he went into another reversal, perhaps the most potentially lethal crisis yet: a gradually developing edema—internal swelling, as body tissues, awash with waste secretions, finally became saturated—that rapidly turned massive. Parts of his body swelled up grotesquely: arms; legs; his scrotum, which ballooned to the size of a football. Under this intense pressure, Johnson blacked out, burning with fever. Suddenly he was again as close to death as he'd ever been.

His body had been so deprived of life-sustaining elements, especially nitrogen, by the long outflow of protein that drastic—and possibly dangerous—steps now had to be taken. His forced feeding was tripled: intravenously with a new serum of albumin (a natural protein substance containing many vital elements), pure nitrogen, amino acids, and casein (a phosphoprotein found in milk products); and by stomach tube with a high-caloric, high-vitamin concentrate of brewer's yeast, vitamins A and D, ascorbic acid (the C vitamin found in fresh fruit and vegetables), thiamin, and riboflavin. The serum albumin, a recently developed substance barely out of the testing stage, was meant to control the viscosity of the thinned, watery blood and regulate the uninhibited flow of fluids within his body; and the compounded diet was intended to give his metabolic processes enough nourishment to continue functioning. The danger was that his system would not be able to absorb such an influx of rich elements all at once and would go into final shock. But that risk had to be run; if *this* didn't work, it was the end for him in any event—for sure, this time.

Within twenty-four hours—once more to the staff's astonishment—the edema was halted. And it began noticeably to ebb. Moreover, tests over the next few days disclosed that for the first time his body was beginning to retain more of the infused protein than he was losing. Johnson regained consciousness and said the pain was less than it had been. He even looked better—had new color in his face, was more alert. His weight leveled off at 110.

The concentrated infusions were continued for several more weeks. By then the edema was completely reversed. And in that time the "tanning" of his wounds also began to take; that porous outer shell at last was hardening, keeping in the draining fluids.

Incredible as it seemed—as it *was*—the gritty (and perhaps supernaturally blessed) young Coast Guardsman had licked death once more. He was going to live. It really was rather like a miracle.

When the word got out, there was jubilation in Boston, almost as if the Red Sox had just clinched the pennant.

And in Sumner, Missouri, joyful thanks.

At about that time, Grove bartender Bill Shea expired of his burns, bringing the death total to 489.

One of the great concerns for many Grove victims and their families was the prospect of impossible hospital and medical bills to be paid. This was particularly true for those who had been delivered to the private hospitals—most notably Mass General—that most could not afford and would not have chosen if they'd had any option. Recognizing this, Mass General director Nathaniel Faxon, for one, early on had made an announcement that the elite hospital would not tender bills to any disaster victims nor would its doctors charge for professional services: "It is a community misfortune, and its toll should be paid by the community at large."

At City Hospital, of course, it was the practice, indeed the purpose, to afford medical care in all circumstances to all in need regardless of ability to pay; but as City had received by far the greatest number of Grove victims, it was all but overwhelmed by the extraordinary requirements of funding the expenditures accumulated during and in the weeks and months following the emergency. To this end, the Red Cross provided considerable relief in the form of financial aid to both the public and private hospitals and, particularly in the case of City, with generous volunteer medical assistance.

Still, to the community—the city and the state—fell the bulk of responsibility for making up the enormous costs. Nobody would ever calculate precisely how much the Cocoanut Grove fire did cost in terms of medical bills, or how long it took to pay off the debt—there were just too many diverse, complex, unaccountable factors—but the most knowledgeable estimates ranged from "over $1 million" to $1.5 million. And that would have been just hospital expenses. Doctors' bills, had they been rendered, would have been enormous.

There was also the matter of reparations to the families of both the dead and the injured. What was the value of a life destroyed? A crippled body? A permanently scarred psyche? Many cringed to put a dollar value on such damage. Yet, to be practical, there *were* crushing costs that would have to be met: funeral and burial expenses, extended private medical care, to say nothing of irretrievable loss of income. And there was also another, perhaps even more compelling consideration, bitterly shared by the victims and their families: the Cocoanut Grove—the *people* who'd operated it and were seen as most clearly liable for this outrage—somehow should be made to pay.

Thus, in the first months after the fire more than five hundred individual claims had been filed for damages totaling over $8 million.

But the claimants soon found themselves frustrated. For they learned that the Grove had not been required by law to carry liability insurance on its patrons—and therefore it had not. (Nor had the management covered its own employees against accident under workmen's compensation insurance, which, though available for a decade, was not compulsory in Massachusetts.) Moreover, fire insurance on the establishment—spread for some unfathomable purpose among six different carriers—was found to be woefully inadequate, in fact hardly more than token: about $22,000.

Within days of the fire, the United States district court sitting in Boston had appointed as receiver and trustee in bankruptcy for the defunct Cocoanut Grove a nationally renowned Boston attorney named Lee Friedman, who specialized in bankruptcies and corporate reorganizations. Friedman in turn had assigned one of his firm's bright young attorneys, Frank Shapiro, as his full-time counsel in disposition of the Grove's affairs. (Shapiro, as it happened, had himself been touched closely by the fire. That Saturday night, he and his wife and some friends had planned to go to the Grove, but they couldn't get in and had strolled over to the Metropolitan Theater instead. Coming out later, attracted by the sirens, they'd gone to witness the awful tumult around the blazing club. Later, to his horror, Shapiro learned that a cousin of his, Dr. Joseph Dreyfus, and his wife had been among those trapped inside; the doctor survived, though terribly burned, but his wife died in the blaze.)

One of the first problems Friedman and Shapiro encountered was in trying to unravel the tangled threads of the Grove's ownership. Without establishing clear ownership, it might be exceedingly difficult to fix legal responsibility for the administration of damage suits. Though Barnet Welansky had been known as "owner" since 1933, not until months before the fire had his name actually appeared on the annual corporate registration. For the first couple of years after Welansky had acquired the Grove from the estate of King Solomon, Angelo Lippi had been nominal "president and treasurer."

Then in 1935 the title had been shifted to Benjamin Welansky, a younger brother of Barney's, and he had been so listed until 1942. When Benjamin entered the Army, Barney himself finally named himself president-treasurer—but only of the specific nightclub enterprise; the *property* on which it stood, at Nos. 13, 15, and 17 Piedmont Street, was apparently registered independently in the name of one of Barney's sisters, Jennie Welansky.

The 1942 city assessment of that property was for $37,800; the club itself was assessed at $22,200. The New Cocoanut Grove Corporation had listed its total assets at $122,180. None of this, of course, would make more than the barest dent in the $8 million in claims. Moreover, to secure so much as a pittance for each claimant, legal possession of the site would have to be acquired by the receivers. But at that juncture, Jennie Welansky did not seem inclined to relinquish such title.

The early outlook, then, for any real compensation appeared bleak, if not completely hopeless.

Still, as receivers the attorneys did have exclusive, court-mandated jurisdiction over anything that remained on the Grove premises. And shortly they got one break.

Fire inspectors continuing to sift the ruins for clues to the fire's origin came upon an interior wall downstairs, in a recess of what had been the Melody Lounge, that was still intact, hardly even scorched in fact. There seemed to be a hollow space the other side of it. Following orders, the inspectors called Frank Shapiro for clearance before investigating further. He gave them approval to rip it out.

It was a false wall. Behind it—and in other hidden places subsequently uncovered in the basement—they found a huge cache of spirits and imported wines: in all, more than four thousand cases of assorted liquors undeclared on any inventory statement, and none bearing federal tax stamps!

When Shapiro and Friedman proposed auctioning this unexpected treasure at the best possible price, for the benefit of all the claimants, they were hit by flak from two sides in Washington: the Office of Price Administration (OPA), a new bureau empowered to stabilize and enforce consumer price ceilings in the wartime emergency, insisted that the liquor could *not* be sold at auction but only within its set price limits, which were much lower than the free bidding would undoubtedly bring; and the Internal Revenue Service demanded, in any case, payment of the required duty—which, if computed by normal guidelines, would take another huge bite out of the proceeds.

The first thing Friedman did was to take the OPA to federal court. In arguing against their position, he cited both the extraordinary circumstances and his own primary authority as a federally appointed receiver. The court upheld him. Then he and his deputy, Shapiro, went to Washington, where, with the vigorous support of the Massachusetts congressional delegation, they appealed to the Internal Revenue commissioners to forgive the usual taxes in this case out of consideration for the horrendously damaged victims of the Grove. The IRS compromised: it would not "forgive," but it would grant a reduction of duties to an acceptable minimum. Friedman borrowed $15,000 to pay the tax bill at once so as to avoid any accrual of interest.

The receivers sold the liquor at auction in Boston, disposing of the entire stock to three hotels—the major buyer was the Parker House—for $171,000. It was a start, at least, at collecting something for the surviving families—even though if shared equally it would come to about $300 apiece at best. (Death benefits from Social Security also would be negligible: from a low of $20 to a maximum of $85 a month.)

Then they got another jolt: on March 3, 1943, Barney Welansky—less than two weeks before he was to stand trial in Massachusetts for manslaughter—was indicted by a federal grand jury on charges of evasion of income taxes and excess profits taxes dating as far back as 1936.

If the government won that case, and exacted every penny

Welansky owed—which his family might have to scrape up by selling the Grove real estate—there would be *no* further source of reparations to the victims.

One problem facing the prosecutors in suing for manslaughter was the important technicality of naming specific individuals who could be *proved* to have been in the Cocoanut Grove at the time of the fire and whose deaths had been a direct result thereof. Notwithstanding the great number of dead, this was not so simple as it might have seemed. In the utter confusion of that Saturday night, hundreds of bodies had literally been dumped at hospitals and mortuaries, many lacking personal identification, much less any incontrovertible proof that they'd been *in* the fire. From a strict legal standpoint it might be held that any of these could have been hurt outside the club, or for that matter could have suffered their injuries somewhere else entirely. So there had to be certainty in order to make the manslaughter charge stick.

Choosing with special care, then, the prosecutors named fifteen who could be verified as having died in or because of having been in one section or another of the club at the time. There were six males (including two youthful servicemen, a Navy ensign and Army lieutenant), and nine females (ranging in age from young singles to older matrons), and among them all one married couple. These figured to blanket the full range of a jury's emotions.

Selection of jurors began on March 15; a panel was seated within two days, and the trial began on Thursday, the 18th. Defending Barney Welansky was his erstwhile partner, Herbert Callahan; counsel for Jimmy Welansky was Daniel Gallagher; for wine steward Jacob Goldfine, Abraham Webber. (There had been speculation that defense counsel would move for a change of venue, on the logical grounds that their clients could hardly expect an impartial judgment in Boston, and some surprise when such a motion was not pursued. It had been considered, indeed, but in the circumstances an inescapable conclusion had been reached: there probably was *no-*

where else—certainly not in Massachusetts—that the tragedy of the Cocoanut Grove was any the less public knowledge.)

The prosecution's witnesses and their testimony essentially were the same as had appeared at the fire commissioner's earlier hearings. Now, of course, there were strict rules of evidence to be observed, and the defense could cross-examine. But few of the incriminating accounts of laxity, probable collusion, and even venality with respect to circumvention of safety codes at the Grove, could be altered or mitigated.

And it soon became evident as the trial progressed that notwithstanding any or all others who might have had a dirty hand in it, the primary liability for the unforgivable deaths of so many could be narrowed down to one individual: Barney Welansky.

chapter 16

Barney Welansky was a seasoned product of the streets of Boston's West End—third-youngest of nine children (five boys, four girls) in a lower-middle-class family, born of necessity with an instinct for survival.

As a school kid he'd scrambled for every penny, working at odd jobs—among them hawking newspapers. By adolescence, he'd so impressed a local publisher's agent with his unflagging hustle that he was handed his own newsstand operation at the busy intersection of Massachusetts Avenue and Boylston Street. Barney put himself through Boston University and then the BU Law School, where he received his law degree when only twenty-one. Too young to practice, he went another year for his master's degree and was admitted to the bar in 1919.

A naturally canny, businesslike attorney with an unsentimental eye for detail and legal fine print, Barney's career had, nonetheless, been unimpressive until he teamed up with Herbert Callahan. The more flamboyant Callahan stalked the criminal side of the law, and the occasionally juicy fees to be derived from that sort of trade; with Welansky's interest more on the cut-and-dried business side, they complemented one another neatly. And they had done well together even before being awarded the joint retainer from Charles "King" Solomon—whose accounts on either side of the law were enough to guarantee both considerable activity and security to the partners.

One of Solomon's ostensibly legitimate enterprises, and clearly his favorite, was the Cocoanut Grove, and after Welansky deftly engineered its distress purchase from the Alperts in 1930 Solomon entrusted to him the responsibility of keeping a sharp eye on its operation—including keeping Solomon's name out of any legally traceable connection with the place. Barney was not, at that stage, particularly enamored of the restaurant or cabaret business; nor, as an attorney in good standing, was he comfortable with Solomon's introduction of illegal liquor there. But above all he was pragmatic; he could adjust to discomfort. And after a couple of years of roaring success for the club he had to concede a certain excitement with it all.

Then, suddenly, that January morning in 1933, the King had been blown away in the men's room of the Cotton Club, and Welansky and Callahan's lucrative retainer with him.

Solomon's primary interests would no doubt be picked up by his associates, but what of his number-one pastime, which he had orchestrated so personally (and at such questionable expense)—the Grove? Who among them had the qualifications, or the desire, to keep the place running? Maybe, Solomon's successors debated, they should just shut it down and write it off as a no longer affordable extravagance.

But there was one who felt the Grove had too much going for it to write it off, who believed that, with the application of some basic cost controls, it could become a profitable ongoing enterprise. That one was Barney Welansky, and he proposed to relieve the others of the burden: he would run it himself.

Solomon had left a widow, Bertha, who had little interest in and less knowledge of the restaurant business; and Welansky managed, apparently without great difficulty, to persuade her to yield any claim to proprietorship of the club. (Just *how* he brought this off—by what settlement offer, if any, or on what legal terms—intrigued observers for years to come. And none ever managed to satisfy his suspicions. Most preferred to believe that Welansky had either coerced or bamboozled Mrs. Solomon out of the ownership that was rightfully hers.*)

Having gained control, Welansky began pulling some legal strings. He declared, for purposes of reorganization, that the club was bankrupt. Next he took out a chattel mortgage on the property, with himself—as its executor—the guarantor. Then, after several months of nonpayment (to himself!), he foreclosed the mortgage for "default." Whereupon, in a foreclosure sale, he bought the place back for a mere $3,000.

But when the reorganized enterprise was registered with the city as the "New Cocoanut Grove, Inc.," Welansky was *not* listed as owner but only as one of three shareholders, the others being Angelo Lippi, the maître d', and one Katherine Welch (not otherwise identified, but who was in fact a minor employee in Welansky's law firm). Lippi, furthermore, retained the titles bestowed on him by the club's previous owner: "president and treasurer."

For all these manipulative exertions, Welansky really did not have in mind *dedicating* himself to the Grove; it was distinctly secondary to his law practice. He would set guidelines for its profitable operation, and Angelo Lippi, under his mantle, would carry them out. His first major change was a dictum that the "star" entertainment policy initiated and nurtured by King Solomon—which had brought the Grove international celebrity, but was in cold reality prohibitively expensive— would be no more. The Grove had *made* its name, proposed Welansky; now, on the strength of that established appeal, still with its exotic atmosphere, fancy service, and acceptable food, it should run sufficiently well offering only simple, good

---

* Such skepticism seemed reinforced when, some time after her husband's murder, Bertha Solomon filed in probate court a formal accounting of his estate. Among "assets" was listed: "Sale of Cocoanut Grove, Inc. (*no value*)." And among expenditures: "B. Welansky, legal services—$500."

dance music and maybe an attractive singer or two. Angelo Lippi, the professional, questioned this; but Welansky was adamant.

Time, Welansky felt, was on his side. For with the passage by Congress of the Twenty-first Amendment to the Constitution repealing the Eighteenth—Prohibition—the Grove, able to sell alcohol legally, would not *need* high-priced entertainment. Trust my instinct, he assured Lippi.

Repeal was ratified into law on December 5, 1933, making most of America legally "wet" again. Before noon that day, carpenters were at work constructing the Cocoanut Grove's first bar. To secure a table there for that historic night, reservations were a must—many were booked on contingency weeks in advance by regular customers—and soon after the doors opened at 7:00 P.M. the club was filled to capacity with a holiday-spirited crowd.

Five new bartenders mixed and poured drinks behind makeshift serving counters as the carpenters even then kept hammering away, putting the final touches on the long permanent bar off the dining room. Just before 9:00 P.M. the construction was completed. A sustained cheer went up from the assemblage as the bartenders moved their paraphernalia to the sleek new location.

Patrons began to call out for some sort of commemorative ceremony: "Break out the champagne!" "Speech!"

Had King Solomon been presiding at such a time, he surely would have risen to the occasion—perhaps even climbed atop the bar and led a pep rally. Barney Welansky was hardly that type; he sat back as usual at a corner table taking it all in, pleased but withdrawn. The crowd instead sought out the one figure familiar to all who'd frequented the Grove: Angelo Lippi. The elegantly continental maître d' customarily disdained any public demonstration, but now he could not shake off the insistently festive throng. With a nod of approval from Welansky, Lippi mounted the stage and waited for attention. In his finely tailored dress suit, with trim black mustache and pomaded hair, he looked indeed like the "Count" he'd been respectfully nicknamed. When they hushed, he raised his hands:

"My dear ladies and gentlemen. It is my sublime pleasure to inform you... *the bar is open!*"

The roar that went up rattled the palm trees.

That was a most memorable night—in fact, however, *the* most memorable at the Grove for quite some time. After so auspicious a beginning, Barney Welansky would oversee a long, steady decline in the club's popularity.

Now there was no lack of liquor to be blamed. Every place could serve liquor now, and that was part of the problem: under Welansky's prosaic cost-conscious rule, the Grove seemed to have lost much of its uniqueness, its verve and "glamour"; and at the prices still being charged, people came to feel they could find more entertainment value for their money (which was growing tighter as the gloom of the Depression deepened) elsewhere.

In general it was a tough, competitive time for luxury enterprises such as nightclubs—and the more so for clubs that offered nothing "special." Which was the state to which Welansky, within less than two years, had brought the Cocoanut Grove. He did not even have that intangible asset enjoyed by its former owners: the "personal touch," that elusive proprietorial charisma that could attract a loyal following. Welansky's absentee ownership became a tangible debit.

Actual management of the Grove was left largely to Lippi, the latest chef (Louis of Paris and His French Cooks were long gone), and the resident bookkeeper, Rose Ponzi. But under Welansky's strict budgetary controls, which Rose dutifully enforced, there was no way to keep the club up to its former standards. The entertainment was limited to a rather ordinary dance orchestra; the food devolved from pseudo *haute cuisine* to standard, or less, nightclub *ordinaire*; service deteriorated from crisply continental to desultory, even careless. Even the vaunted Grove "atmosphere" itself showed telltale signs of increasing shabbiness and disrepair. And as business diminished accordingly, management's response was only to cut back staff and services yet further, which inevitably magnified the problems still more.

Early in 1935, the ineffable Lippi himself finally lost what hope he'd clung to and tendered his resignation; he'd accepted an attractive offer from the quietly swank Somerset Hotel. Welansky could not deny him the more promising opportunity—as, say, King Solomon might have done—and indeed wished him well.

(The maître d'hôtel had been settled at the Somerset for several weeks before he was reminded that, though departed, he was *still* carried on the Grove's books as its "president and treasurer": that was when Rose Ponzi came to the hotel with a sheaf of payroll checks for him to sign. Welansky was still playing his little corporate games! Lippi knew what some people were saying—that Welansky was bleeding the Grove as some kind of arcane tax dodge—but he couldn't believe that, did not *want* to believe it. Bleeding? The place was already in the red. But then Lippi didn't understand finances, just as Welansky did not understand nightclub operation. *That* was the problem: the man had meant well, Lippi was sure, but he just hadn't known how. It was a shame.)

Barney Welansky's intent had *not* been to steer the Grove into liquidating. He had applied what he'd thought were sound business practices toward making it a viable, self-sustaining enterprise, but even he could now see that his design had not worked. Well, he could always put it up for sale and get out without too much damage, maybe even make a modest profit. After all, it *was* the Cocoanut Grove—surely there would be no dearth of interested buyers.

So, in 1935, discreetly he let it be known around Boston that the Grove might be had for as little as $45,000, free and clear—and shortly began to realize he'd been wrong. There was not a nibble, through all of that year. By then Welansky— spending more and more of his valuable time running the place himself with a skeleton staff, and watching the losses mount—recognized that the grim national economy had turned *against* him. The Cocoanut Grove was looking more and more like a white elephant.

He thus found himself with just two options: let it collapse of its own weight and use every wile he knew to cut his losses; or bow his neck, climb on board with both feet, and give it a real run for the money.

Perhaps it was the redoubtable Angelo Lippi's defection that had at last stirred Welansky. Or perhaps it was simply that he was a prideful man with a fiercely competitive family background. In any event, late in 1935 he decided he would not *accept* failure; he would reverse his life—become an entrepreneur first, a lawyer second—and put his every effort into vindicating his decision.

That October, when it came time to renew the Grove's license, Lippi of the Somerset found he'd been quietly "retired" as president and treasurer of the New Cocoanut Grove, Inc. His successor of record was Barney's brother, Benjamin Welansky. At least they were keeping it in the family, noted the bemused Lippi.

From then on, Barney did run the Grove almost full-time, virtually disassociating himself from most other interests, even his legal practice. (He continued to keep a law office with Callahan, up to the very end, but in the last years it constituted no more than a mail drop, telephone, and a part-time secretary.) Thus commenced the Grove's second notable "comeback"—the beginning of its third great era in what would be, after all, a relatively brief history of only fifteen years.

As Welansky had systematically run it downhill, he now turned it around and—though not so easily—started it back up. He finally put some money into the place, mostly his own now. First, physically, it got a going-over, some badly needed painting here, cleaning or mending or even replacement of worn furnishings there, a general sprucing of the cocoanut-palm motif. He shook up the kitchen and service staffs, actually *adding* experienced help rather than settling for less, or eliminating, to save a dollar. He gave more attention to the entertainment lineup, calling on local booking agents to find the best available talent (if still at the most reasonable price). A renewed life, a discernible zip, gradually was infused into the place.

Gradually, word got around that the Grove had picked up, and people started drifting back. To see people, bodies, again filling the club became an obsessive goal for Welansky. He put on special "nights"—celebrity nights, amateur nights, "nation" nights—anything short of giving away dishes, to bring them in. He didn't roll back his menu or drink prices, in the sound conviction that people who went out *looking* for a good time didn't much balk at paying when they found it. And they came, and he charged them, and they seemed to enjoy, and they paid.

With affairs firmly on the upswing, Welansky went to the Somerset Hotel and talked Angelo Lippi into returning once more to "the place he belonged." That was the symbolic turning point.

The economy had begun to turn as well, to the advantage of escapist service businesses like the Grove. By 1938, while the Depression was not over, there were at last signs of light from what could be seen as the end of that seemingly endless tunnel. The threat of another great war was spreading fearsomely across the world, and nations, including America, were beginning to gear up for rearmament. That meant factories reopening, manufacturing and trade gradually picking up, and more people going back to work and once again taking home paychecks (generally fatter now, thanks to the trade unions, than ever before)—all of which meant there was more money to be spent.

So by 1938 the Grove had regained much of its lost favor in Boston—and *without* any concession by Welansky to the old "big-name" show policy. He had been able to merchandise a consistently reliable entertainment package consisting of dance music and floor shows that changed every few weeks and featured a variety of fresh young performers who though not yet stars might well be in the future. In short, the Grove had become an exhilarating showcase for new talent.

Indeed, things were so looking up that Welansky was emboldened to expand. Years before, Solomon had purchased the buildings to the east on Piedmont, between the Grove and Broadway, and had them razed, no doubt in anticipation of expansion; but what plans he may have had, of course, had been cut short, and the empty lots that remained had been used as a parking lot for the club. Maybe one day, Welansky thought, he *would* build there; but not just yet. Instead, when the opportunity presented itself, he bought the small building abutting the Grove to the west, the basement of which was adjacent to his kitchen and storage areas. And down there he had Reuben Bodenhorn design a bistro—a room with atmosphere suggestive of the exotic motif of the main club, but intended for those who wanted to escape the hubbub and glitter, a dark and romantic hideaway for lovers. He had the perfect name for it, casually inviting: the Melody Lounge.

The new addition was an immediate hit. It seemed that for Welansky *everything* was coming up roses.

If there was one aspect of his flourishing new setup that Welansky had not felt entirely easy about, it was the absence still of a standout "personality"—someone with sparkle and wit and show business savvy, who could not only package the classiest shows in town but *present* them out front, act the host and charm the socks off the people and make them come back for more—some pro with a special brand of pizzazz, who would be Mr. Cocoanut Grove. It was not a role Barney himself was ever suited for, or one he cared to assume. But nobody else he had quite filled the bill.

It was about then that Mickey Alpert came back to Boston. And after that, things only got better, faster.

Barney and Mickey, though very different in many ways, found a kinship in others. Both were tough-minded, knowing what they wanted and how to get it. Barney never concerned himself with fostering affection or even admiration among those he employed; he insisted only on a good day's work for a day's pay. And Mickey, as outgoing and irrepressible as he was before an audience, offstage was himself something of a martinet—not greatly admired, much less loved, by most who worked for him or whose employment depended on his approval. Out of the spotlight, he tended to switch off the warm stage glow; his manner became cool or even brusque toward the musicians and performers, all business or sometimes downright unfeeling. That Mickey was all show and no talent became a smirking byword among his supporting cast.

He and Welansky also shared a tightfistedness with the dollar. And because of that, one or the other of them—or both—finally fomented an uprising among members of the orchestra early in 1940.

Most employees at the club were paid individually by the bookkeeper, Rose Ponzi, but Mickey collected salaries for the entire orchestra and paid the musicians directly. Scale for union members was then $50 a week (that was all Mickey was paying; if any didn't like it they could just move on); but for some time each man found he was averaging only about $35. Mutterings began. Were their salaries being "skimmed"? By whom? Mickey? Or by the band member—designated personally by Mickey—who acted the part of "contractor" between the club and the local union?

A delegation of them went to Local 9 to complain. But there they were rebuffed by edgy union officials who cautioned them gravely that in challenging the management of the Grove they might be messing with the wrong kind of people, leaving the indelible impression that Welansky & Co. (Mickey Alpert presumably included), if pushed, might just take recourse in some unpleasant self-defense. Not only was that revolt abruptly shot down, but as a result the whole orchestra got itself fired, and Alpert put together a new ensemble from scratch.

Then, only the following year, a new dispute arose over the musicians' having to work overtime without extra pay.

This had to do specifically with Mickey's big Thursday "celebrity guest nights" (which he had soon built from sometime events into a regular weekly attraction): normally the band was contracted to work until closing at 1 A.M., but on Thursdays they were often required to play well past that hour, to 2:00 or even 3:00, yet this was never reflected in their paychecks. Accrued back pay had accumulated to as much as $150 a man when again they appealed to the union.

This time Local 9 stuck to its guns, demanding the Grove pay up or face a boycott.

Barney Welansky was irate, and Alpert embarrassed at having been shown up. In the circumstances there was little they could do by way of mass retaliation... but Welansky (with or without Mickey's acquiescence) did find a way to make a point: when the orchestra's contract came up for renewal, as was customary at each year's end, all were rehired but one—the one management had fingered as instigator of all the fuss.*

So, all was not sweetness and light at the Grove even in those booming times, but nobody wanted to leave: the place was on a roll, jam-packed night after night, the stardust was back—they all knew they were riding a winner.

Until it all came crashing down around them that Saturday night.

The manslaughter trial of Barney Welansky—for that's what it was, really; his brother and Jacob Goldfine were almost

---

* The "one" was bassist Jack Lesberg—who, actually among the most easygoing, was anything but a troublemaker. In fact, Lesberg had been one of the few Grove musicians who got on reasonably well with Alpert; nor had he ever expressed ill will toward Welansky. It simply happened that Lesberg had close personal ties to the union executive committee (through his brother, another locally well-known musician), and had been asked by his fellow players to transmit their complaints. Sacked, Lesberg went off to join the Muggsy Spanier band. Then, interestingly enough, in November 1942, when Alpert needed a new bassist at the Grove, it was Lesberg he asked to come back.

incidental—lasted three weeks. Nobody seemed seriously able to consider either the hapless Goldfine or Jimmy Welansky as an accused. (One point brought out against Jimmy even proved somehow to reflect more damagingly on Barney. Just four months prior to the Cocoanut Grove disaster, on July 29, 1942, a fire in an artificial palm tree at Jimmy Welansky's Rio Casino—just blocks from the Grove—had routed some two hundred patrons. Fortunately the blaze had been quickly extinguished and no one was hurt. But shouldn't that incident have provided a lesson?)

Barney, in his defense, denied having knowingly circumvented, flouted, or broken city safety codes, or having invoked his alleged "influence" at City Hall or any other city department to curry special indulgence; he flatly denied having made or offered any "payoffs." He tried to explain the barred or concealed exits and other mean of egress from the club: the Shawmut Street door to the building housing the new dressing rooms was locked in the evenings to protect the performers inside, most of them women; the others, in the club itself— well, they were shut off mostly to prevent too much coming and going... frankly, to discourage deadbeats from running out; it had been the practice for years, not any diabolical scheme he'd just dreamed up. Fire and police and building inspectors were around all the time; nobody had ever said anything, and he'd never had cause to give it much thought. The fire, all the lives lost—no one could have anticipated such a terrible thing. It could have happened anywhere... he wished to God he could bring them all back, but... how could *he* have known?

In the courtroom, pulling for both his uncles, and grieving particularly for Barney, was Daniel Weiss, who since the fire had begun his internship at Boston City Hospital. He was *Dr.* Weiss now, thanks in large degree to Barney's encouragement and support. Barney was a limp, pathetic figure, so tired and clearly not a well man. His defense had been lame, flat. Yet Weiss believed wholly in his essential blamelessness. How could *anyone* be "guilty" of such a crime as the one with which he'd been charged? But Weiss sensed the ominous mood in that court, among the jury. *Somebody* had to pay....

(While the trial was in progress, there occurred another interesting development outside the courtroom independent of what was emerging there, but perhaps reflective of a basic issue: rampant dereliction of duty, if not outright corruption, among city officials. Boston Police Commissioner Joseph Timulty and six of his chief subordinates, including Superintendent Edward Fallon, were indicted by another Suffolk County grand jury for conspiracy to permit operation of illicit gambling houses and bookmaking establishments in the city. All pleaded not guilty and were released on bail. However, Governor Leverett Saltonstall called the seven to the State House, where Timulty—who had been appointed seven years before by James Michael Curley himself—was stripped of his rank and the others summarily suspended from duty. A captain, Thomas Kavanaugh, was named acting commissioner.)

On Saturday, April 10, the manslaughter trial went to the jury.

The verdicts were returned in less than five hours:

Jacob Goldfine: *acquitted* on each count of the indictment.

James Welansky: *acquitted* on each count.

Barnet Welansky: *guilty* on all counts.

On April 15, 1943, Welansky was sentenced: twelve to fifteen years on each count, to be served concurrently, at "hard labor" in the state prison at nearby Charlestown.

The judge asked him if he had anything to say. Welansky—sallow, shrunken, appearing many years older than forty-seven—stood mute.

At the gray, dank monolith of a jail across the Charles River, he was assigned a sewing machine to make inmates' underwear.

chapter 17

In mid-January, the six special nurses provided by the Red Cross to care for Clifford Johnson had begun to withdraw, either to take up other assignments or for personal reasons. Several were close to exhaustion, both physically and psychologically, and simply could not continue. It was a terribly demoralizing job, working all those weeks—two each on three eight-hour shifts—tending that horribly disfigured boy who should have died, would have been better off dying, but would not. The emotional drain combined with the sheer arduousness of their duties created a strain that few individuals could have withstood indefinitely without cracking.

And now, just when Johnson, miraculously, was showing faint signs of life, they reached the end of their own endurance and one by one begged to be relieved. It created an added problem for Dr. Newton Browder, critical in its own way. Regular hospital personnel were too busy to provide constant care needed by the pitifully helpless patient, and student nurses would not do. Philip Butler attended Johnson as often as he could, but he, too, had other duties. Browder, determined not to forfeit this remarkable case through any avoidable circumstance—and much as he may have wanted to, he could not himself spend *all* his time on Johnson—calculated that two seasoned nurses, along with Philip Butler, might just suffice. It would be tough on them, one on duty all day, the other all night, but if he could find the right ones, it might work. Browder again went to the Red Cross and begged them to locate such a pair.

It took time, but the local chapter finally came back to Browder with two whose credentials seemed to meet all of his qualifications. Both were semiretired, with strong hospital backgrounds, and both had recently completed voluntary tours of duty with other Grove victims.

One was a spinsterish, feisty little woman of thirty-seven, named Mercy Smith, short-spoken but crisply professional. She was the first to come in, and Browder assigned her, for her greater experience, to the busier day shift.

The other, Ellie Kampper, ten years younger, had given up regular nursing when she married a few years earlier, but returned to it after her photographer husband went into the Navy. She would be Johnson's night nurse. (One of Ellie Kampper's previous Grove patients had been the young singer, Dorothy Myles, whose once-lovely face and hands had been so disfigured. Ellie had grown quite close to the girl—so sweet and innocent and fearful that her performing career was ended—and it broke her heart to think of all the agonizing surgery Dorothy would have to endure before she could ever hope to be whole again.)

Smith and Kampper, along with Browder and Butler (and City's chief of burn research, Dr. Lund, consulting), thenceforth would be Clifford Johnson's sole links to the living world.

It was this faithful team that had attended the sailor when he underwent and overcame the edema crisis in February, giving them the first real sign that he might actually, by God, pull off the miracle and recover.

Newton Browder and Charles Lund and everybody else at City Hospital—including Robert Aldrich, the burn expert from Johns Hopkins, whose triple-dye refinement of the laborious tanning therapy had become the standard burn treatment there—were, of course, well aware of the signal success enjoyed by Mass General in its use of Oliver Cope's simplified new approach. It was the talk of the medical world, and indeed City and hospitals everywhere already were beginning to adopt some of Cope's methods.

But City, in the thick of the late emergency, had not been prepared to follow other than previously accepted procedure. The doctors there knew how long and sometimes inefficient a method it was, but it was all they had to call upon at the time. The tanning process had very nearly failed for Clifford, and Newton Browder had to wonder whether Cope's treatment would have been more effective, whether the boy would have responded sooner, more positively, to Mass General's new methods. It was a question he could not answer, but one thing he did know: Mass General had not had any patient with burn damage anywhere near so extensive as that of Clifford Johnson, and he sincerely hoped it never would.

Something else Browder knew, and could be proud of, was that he and the good, dedicated people at City Hospital had blazed their own path in keeping Johnson alive. Before the Grove, it had been considered virtually medical anathema to introduce more than a minimum of fluids into a patient suffering extensive deep burns. But they had learned, by sheer necessity, that this was exactly the treatment necessary to forestall shock and death. Burns discharged vital body fluids; they had to be replaced—it seemed so plain now; how could they not have grasped it sooner?

Still, for all this, Browder could not help wondering how much their ministrations had actually contributed to Johnson's survival, and how much was attributable to the lad's own strength and determination. Was there some subconscious, innate *will power* in human beings (or in some, anyway) that could take over even without its possessor's conscious knowledge? It was another mystery the doctors might never be able to solve.

Nonetheless, it was far from time yet for either self-congratulation or rumination. Johnson was hardly out of the woods. While he appeared to recover strength each day, his net weight gain was still negligible, and his frame remained skeletal. And while for the most part his worst wounds definitely were beginning to heal, no longer oozing sera as before, the entire posterior of his body—from upper neck and shoulders to the soles of his feet, and the undersides of both arms—was still a mass of dead tissue, black and coarse as charcoal. That soon would have to be attended to.

Coast Guard buddies lately had been permitted to visit Johnson, whose nurses, with amused understanding, noted their invariable reactions. The initial response was always the same: robust, buoyant young men would abruptly become subdued, even timid, as they tiptoed into the room; all they could see at first was the blackened, shapeless form stretched out length-wise, facing away from them. Tense with wonder, they would edge around toward the head of the bed... and it was only then, seeing Clifford's white, unmarked face peering out of that forbidding hulk, that they would understand what they were seeing—and almost without exception they would blanch, gasp or swear weakly, turn gray with nausea. The nurses understood, because even for them, in all the hours and days and weeks they'd been in this room, it never was easy to get used to that sight.

In order to become whole again, or reasonably close to it, Johnson would have to go through yet another, longer, perhaps equally torturous ordeal that would test anew his inner strength. With the extent of his burns, he would face months of delicate skin grafting, with *no* guarantee, even if the rest of his body could supply enough healthy tissue for the vast transplanting necessary, that the grafts would take. He could wind up a horrible freak, scarred and crippled, half a man, for the rest of this life he'd somehow stolen. Nevertheless, it had to be done.

By the middle of March—in Johnson's fifteenth impossible week at City Hospital—Dr. Browder decided it was time to start trying to put him back together.

Normal grafting, or "plastic surgery," for burn patients entails transplanting sizable patches of living skin onto the burn sites.

But Johnson had so little whole skin with which to work that Browder had to go to the more exacting process of pinpoint grafting. In this method, by the delicate use of razor-sharp surgical needles, tiny dots of healthy skin, each no larger than one millimeter (about 1/25th of an inch), are taken one at a time from the "donor sites"—with Johnson, from the unburned outsides of his arms to start—and "planted" in the raw destroyed tissue a centimeter apart so as to leave room for them to "grow" and join. It is a meticulous, slow procedure demanding extraordinary fortitude, not least by the patient. The pain would be exquisite, and Johnson would have to be given not only Novocaine locally but extra codeine as well.

The first session, performed by Dr. Browder, two staff physicians, and two nurses at Johnson's bedside because it was not yet safe to move him, lasted over three hours. More than 1,500 pinpoint grafts were transplanted from his arm to his back before he couldn't stand it any longer.

Browder waited ten days to check his handiwork—and then anxiously, for he knew how great the chances were of things going wrong. It was not unusual to lose many, or most, of the patchwork of microscopic grafts because they had failed to settle properly, or simply because they had stuck to the bandages. But when he gently undid the lubricated dressings, he found that almost *all* the grafts had not only held fast but were already growing—spreading into the beginnings of a new layer of living flesh! Browder started planning the next operation at once.

They did one session a week, three to four hours each time, well into April. And after each operation most of the grafts appeared to "take." To the doctors, the miraculous elements of this Johnson saga seemed to have no limits. At the end of a month, after some six thousand separate transplants, Johnson's back was no longer black but, from head to foot, pink with new growing skin.

It was time to turn him over and do his other side. One day early in May, Dr. Browder had four of the young Coast Guardsmen who'd been visiting Johnson regularly come in together to perform the "ceremony." Doctors and nurses gathered for the event; they watched, hushed, as the sailors positioned themselves at Johnson's feet and shoulders, then, with a "hup," lifted him off his stomach for the first time in five months, carefully turned him over in midair, and laid him on his newly reformed back. All eyes were on Johnson as he settled himself tentatively in this unfamiliar position.

A broad grin spread over the boyish, good-looking face—freshly shaven for this "unveiling"—and he drawled in his Missouri twang: "Gol-darn!" One of his mates let out a whoop, and everybody else broke into applause and whistling.

It was a rare moment for the medical staff—one of jubilation and triumph.

Johnson, withdrawn miraculously (a word even the doctors were no longer hesitant to use) from death and now at least halfway whole, was a bright, garrulous young man. He had become the hospital pet; everybody went to look in on him, ask what they could do for him. And outside the hospital, his fame had spread to the point where fan letters were arriving in bunches, not only from the Boston area but from all over the country—with prayers and cheer and an effusion of offers to send anything he might need or long for. He had mentioned in several replies—courtesy of nurse Ellie Kampper, who many nights found herself acting as his personal secretary—that he loved country (what he called "hillbilly") music and hoped to learn to play the guitar one day; before long not one but two guitars were delivered to the hospital for him.

But Johnson was still quite a way from playing any musical instrument, or even moving very much. The long period of lying immobile on his stomach, spread-eagled, had promoted growths of skin webbing, like that on ducks' feet, in his armpits and the crooks of his arms. These would have to be removed surgically and then he would have to relearn the use of his arms. His legs were also useless, not merely from the atrophy of idleness but because of several infected bone-deep wounds. These conditions, along with another deep wound in his side exposing some ribs, and the burns on the insides of his arms and his hands, were what Dr. Browder would turn to now that he had his patient right side up.

But as the doctors were preparing for this next phase of treatment, they suffered a terrible setback.

Lying on his back, with the fresh grafts there still healing, Johnson soon experienced great irritation—unbearable stinging and itching—and instinctively he writhed in the bed trying to alleviate the maddening discomfort. Browder was not aware of this until, after a few days, he turned Johnson on his side to examine his back. The surgeon recoiled in horror and despair: practically all of the fragile grafts had been dislodged and now were hanging loose stuck to the bandaging!

All the painstaking work on his back was lost. It would have to be done over, from graft one.

Johnson was returned to his stomach. This time there was no "ceremony."

The second trial of Cocoanut Grove accused—for conspiracy to violate the city's building laws—began in superior court on June 15. Jimmy Welansky again was a defendant (represented this time not only by his own counsel, Gallagher, but also by Barney's lawyer, Callahan), along with architect Reuben Bodenhorn, contractors Samuel Rudnick and David Gilbert, and fire inspector Theodore Eldracher.

The trial turned into something of a judicial comedy. Days before testimony was completed, the judge directed the jury to acquit Bodenhorn. Then, when the case was turned over to the jury and they shortly returned verdicts acquitting all but Rudnick, the contractor's attorney exploded in disbelief.

How could his client be found guilty of "conspiracy" when all his previously alleged fellow conspirators had just been declared *not* guilty? Who, then, had Rudnick conspired with?

The judge took this solemnly into consideration—and upheld the verdict (which called for a two-year prison sentence), though he did grant a stay pending appeal.

Rudnick never would serve any time. The appellate division ruled his guilty verdict not improper, but recommended that, in the unusual circumstances, his sentence be suspended indefinitely.

That left the last three of the ten indicted by the grand jury—all city officers charged with willful neglect of duty: Building Commissioner James Mooney, Fire Lieutenant Frank Linney, and Police Captain Joseph Buccigross. It would take a bit longer to bring *them* to a public accounting.

In April 1943, the last hospitalized survivor of the Cocoanut Grove was discharged from Mass General after four and a half months.

In May, the last Grove casualty died at Boston City Hospital after five months—a Dorchester woman who had stubbornly withstood both grave burns and compounded internal injuries until she could resist no more. She was the 490th fatality.

Clifford Johnson's back *could* be regrafted—the unburned parts of his body from which the transplants had been taken were themselves now healed and could be used again—and though the repeated operation would be even more laborious and would make recuperation that much longer, Dr. Browder and his associates set about it with renewed vigor.

When they had repaired the back this time, they would wait longer before turning Johnson over.

Only one further aspect of the case gave Browder serious cause for concern. It was the growing suspicion that his patient might well have developed a dependency on codeine.

From the beginning, Johnson had required regular doses of the narcotic to enable him to live with the searing pain, and during the interminable, excruciating process of grafting and healing he was administered extra dosages. Now, even as his physical recovery had progressed beyond all conceivable hope (by June he'd regained 10 pounds, up to 120, his body was filling out and he was becoming fitter each day), his disposition had changed—and not for the better.

Johnson, the plucky country boy, had turned whiny, unpleasant, demanding more and more of the nurses and attendants. Everything bothered him; nothing anybody did for him was enough; he had to have this or that, and was nasty when he didn't get it or got it late or it was not exactly what he'd said he wanted. Much of this was understandable, considering the tremendous ordeal he'd gone through to get this far and the crushing letdowns. But where before he'd endured his agony, however great, relatively uncomplaining, now more and more he did complain of pain. And he pleaded more and more for codeine. When his nurses could not be moved—though for Mercy Smith and Ellie Kampper his evident need was heartbreaking and they sometimes almost wavered—he would become nasty and abusive.

It was an ominous sign to Dr. Browder, and he realized he had to make a hard decision.

He sat down alone with Johnson and told him he knew what was happening and was going to put a stop to it then and there—however much suffering it would bring. He was taking him off *all* painkilling drugs effective at once. Browder thought the frightened young man might suffer a relapse from dread on the spot. But Johnson, running with sweat, trembling, just gritted his teeth and shut his eyes tight and nodded miserably.

He went through the tortures of withdrawal for five days. And then he awoke one morning calm, his body cool. He even smiled at Mercy Smith for the first time in what seemed weeks and started talking, rambling but controlled, about how much he missed his home in Missouri, the woods and fields, how he could hardly wait to get back to the farm and work with the soil again. Then he dropped peacefully back to sleep, a little smile still on his lips.

The country boy had won another battle. He would never ask for codeine again.

*   *   *

By the middle of that summer, all the grafting and most of the surgical patching of his damaged leg was done. Over four months, more than 25,000 individual dots of skin had been transplanted, and on the surface at least there was hardly any visible sign of the enormous damage he had suffered. The only wound still not mended to Dr. Browder's satisfaction was the exposed leg bone infected with osteomyelitis. The inflammation persisted and made it difficult to close up.

Johnson's appetite had returned with gusto, and as his caloric intake increased so did his weight. He now retained virtually all the protein being fed him—estimated at approximately equivalent to 10 pounds of beefsteak a day—and he'd begun to look actually *healthy*.

But he still could barely move his limbs by himself. Seven months prone and then supine in bed had left his muscles flabby and his joints unbending. (Moreover, the new skin covering more than half of his body had knit stiffly into a tough shell that did not give.) So, even while he grew more robust, he remained effectively a cripple, and the next step would be a long, painful regimen of physical therapy to get him functioning on his own.

First, they had to soften the body surface, restore resilience especially to the hard new skin so that it could absorb the movement of the exercises to follow and not crack or split under stress. This was accomplished, over a couple of weeks' time, through daily massages with cocoa butter. When the skin was fully pliant, they would begin on the motor faculties.

This was harder and would take much longer. Rehabilitating human limbs (and joints, ligaments, tendons, muscles) that had lain unused for so long was grueling work—arduous for the therapists and torturous for the patient. For hours a day, for weeks on end, Johnson's nurses put him through the rigors of making those members work again: pushing, pulling, stretching, flexing, over and over, until he would cry out in agony. The pain for him was almost as great as when he'd lain there, ravaged and helpless, his body screaming its hurt. But now, of course, it was a different kind of pain: now there was *purpose* to it, he had some control over it, he could see the desirable end to it. Now he could put up with anything that would make him well again.

On the last day of August, Johnson got out of bed for the first time since November 28, 1942. Supported by Mercy Smith and another nurse, he shuffled haltingly across his room and back. He was breathless, wet with perspiration, but grinning from ear to ear. He was nearly home! The women, almost as proud, tucked him in, and each kissed him on a cheek.

Two weeks later—after moving farther from the bed and staying up longer every day—he was allowed to walk unaided.

It was a magical moment at Boston City Hospital, like the world premiere of a brilliant new presentation. Radio newsmen, photographers, and reporters from all the papers and wire services came to record the event. Johnson performed for them, a star in the spotlight, walking out of his room, proceeding slowly down the corridor. The press and hospital staff applauded and clamored about him, and, his own eyes wide with delight, boyish face beaming, he could only cry out: "Ain't this somethin'!"

And to everyone there, indeed it was. There was, as they say, hardly a dry eye in the house.

chapter 18

Fire Lieutenant Frank Linney went to trial in October 1943. Held up to scrutiny again was his "good" inspection report on the Grove just eight days before it burned to the ground—an evaluation tragically contradicted by the accounts of witnesses and participants in the subsequent devastation.

After nine days of testimony, the jury deliberated three and a half hours. The verdict: not guilty of willful neglect.

The sixty-two-year-old Linney, who had been impassive throughout the proceedings, finally allowed himself a weary smile when, as he was being congratulated by his wife and friends outside the courtroom, each of the twelve jurors in turn came by to shake his hand and wish him well.

Asked by a reporter if he had any statement to make about his indictment and subsequent acquittal, Linney's smile faded. He paused, as if to weigh his thoughts, then said quietly: "I think maybe I've already made too many statements—don't you?" He did not smile after that.

Building Commissioner James Mooney was up next, early in November. The state's case against him rested on two questions: whether nightclubs were included under the general law that prohibited occupancy of any building used for public assembly until it was issued either a license by the mayor or a certificate by the Building Department; and whether Mooney, as building commissioner, willfully neglected or knowingly failed to enforce such a provision. (The first question caused a great deal of legal confusion, for until the Grove fire the law had never been clear as to whether a "nightclub" did come under the category of "place of public assembly," as did theaters, meeting halls, etc. Only since the fire had the state legislature amended the statutes to confirm the designation.)

The prosecutor, Assistant Attorney General John Walsh, first established that in fact no such license or certificate had been issued to the Cocoanut Grove. Then he read into evidence transcripts of Mooney's testimony before the grand jury in which he had admitted that he had never consulted Boston's corporation counsel for an opinion on whether the law applied to nightclubs.

Mooney's position, in defense, was that he'd always believed licensing of such places to be in the purview of the city's Licensing Board; for the past thirty-nine years, Mooney stated, no building commissioner had *ever* invoked the statutory provision in question. He'd just followed tradition, "the way it had always been done."

Walsh's presentation of the state's case took little more than an hour. Mooney's attorney, former federal judge Hugh McLellan, then argued that, on the evidence, the case should not even go to the jury: all the state had shown, he contended, was that, mistakenly or not, Commissioner Mooney had performed his duty *as he'd seen it*—not with any bad faith or willful neglect as charged; if there had been an error, it was not a matter of *wrongdoing*, but simply of human misunderstanding.

The judge concurred and directed the jury to deliver a not-guilty verdict. After an eleven-month wait, Mooney's acquittal—if not a complete exoneration—had taken under two hours.

The only accused still awaiting judgment then was Police Captain Buccigross, who had been on suspension since his indictment at the end of 1942. Despite the public rancor directed at Buccigross, from a legal standpoint his was a tough case to prosecute, and the attorney general's office had seemed reluctant to pursue it. Among other considerations was the inescapable fact that a Suffolk County assistant district attorney, Garrett Byrne, had been side by side with Buccigross in the Cocoanut Grove's New Lounge that fateful night and had corroborated the captain's account of what happened—both at the Fire Department's public hearings and later under oath before the grand jury. (Byrne had been kept from active participation in his office's pursuit of the case because of his own direct involvement.)

In the meantime, Buccigross and his family had barely survived a year without pay, much less official vindication or personal peace of mind. He had remortgaged their home, borrowed heavily, taken temporary jobs, just eking out an existence until he could be cleared—for he was set on remaining a policeman and being restored to honorable duty. But how long would it take? Could he continue to hold out? And would it prove to have been worth it in the end?

✓ In the year since the disaster, the Massachusetts legislature had introduced and adopted a flurry of new measures regulating fire safety in public buildings. In addition to the act more strictly defining "a place of assembly," another established within the Department of Public Safety a board of standards and appeals to make corrective changes in building inspection and licensing codes. A specific prohibition was placed on the continued use or any further installation of revolving doors as *sole* entrances or exits; henceforth there must be at least one adjacent hinged door—preferably two—which open only *out*. (The revolving doors at the entrance to the State House in Boston were removed soon after.)

✓ Many of the new regulations, in fact, closely followed recommendations set forth in the official report on the Cocoanut Grove calamity submitted by the Boston Fire Department to the state fire marshal late in 1943.

It was an impressive study—sixty-four pages of explanatory and analytical text embellished with statistical charts, diagrams of the nightclub's physical layout, and stunning "after" photographs of the destruction both inside and outside. Appended was a roster of all the witnesses interviewed during the department's seven-week-long public hearings. The report was signed by Commissioner Reilly himself.

Following a descriptive summary of the outbreak of the fire in the Melody Lounge and its astonishly rapid spread throughout the club ("Flame appeared in the street floor lobby within two to four minutes after it was first seen in the basement room, and within five minutes entirely traversed the street floor of the main building and had passed to the entrance to the Broadway Lounge"), the report turned to the causes of its rapid spread:

> Plainly a large and extremely hot volume of burning material, largely gaseous in form.... Much of the cloth, rattan and bamboo contained in the Melody Lounge, and on the sides and lower walls of the stairway leading therefrom, was, in fact, not burned at all, and the same is true of the carpet on the stairway, contrary to all usual fire experience.
>
> ... a major part of the great volume of burning gas projected to the first floor consisted of carbon monoxide gas [which] had arisen as a by-product of the fire, burning with deficiency of oxygen in the low-studded basement room. The cloth false ceiling was tacked to wooden members attached to the underside of reinforced concrete beams in such a manner that there remained a dead space of 16 inches between the actual ceiling and the false ceiling, with a deficiency of oxygen in this dead space. [Therefore] combustion of the cloth was incomplete, and occurred largely on its underside where oxygen was available.
>
> Products of such incomplete combustion... will themselves burn further as soon as additional oxygen is encountered.... In the basement room there was no ready outlet for the heat generated [which] therefore increased

both the temperature and the pressure of the partially burned gases, and acted to drive them forcefully to the nearest available outlet... toward and up the stairway.

...Comparatively narrow [four feet] and rising sharply, the stairway acted like a chimney, adding a draft of suction to the pressure generated in the room below by heat. Such effect appears to have been very considerable, since it drew out the flame entirely, leaving unconsumed the wood and cloth material [below]....

In the stairway itself a further acceleration of the process occurred. Here the partially burned hot gas was rapidly mixed and churned with a considerable volume of air contained in the stairway... further combustion resulting [that] increased the temperature and rapidity of flow of the mass....

The burning mass passed from the top of the stairway into a narrow connecting corridor and thence to the street floor foyer. The wall coverings of the foyer, consisting of artificial leather on cotton batting on concrete, which would be unaffected by ordinary flame such as that from a match, did not withstand this blast of superheated burning gas. The burning and decomposition of such wall coverings once again produced material largely gaseous, capable of further combustion and of very rapid movement....

At this point the only available direction of expansion for the hot, expanding mass was down the length of the foyer... accelerated by a large ventilating exhaust fan placed over the further end of the Caricature Bar [with the] effect of increasing the chimney effect....

The great mass of compressed partially-burned gases spread at once into the main dining room... and into the Broadway Lounge [beyond the Caricature Bar].

...If all the exits had been open, obviously more people would have gotten out of the building alive, and there would have been less retention of gases, heat and fire... but even then many casualties would still have resulted, as fire and persons would still have had to rely upon the same means of egress....

The report next described the extent of Fire Department operations. And it credited the assistance provided by other official agencies and countless volunteers.

Then it examined the "Causes of Loss of Life":

... While it is not clear that the electrical system was completely disrupted, most of the lights on the premises became extinguished immediately upon appearance of the fire. This fact, coupled with [the] smoke and flame and the cries of "fire," produced great confusion....

A considerable number of deaths was caused by the fact that the door opening on Piedmont Street, at the top of the stairway from the Melody Lounge, could not be opened....

Further deaths were caused by the fact that members of the public were unfamiliar with the location of the exits. The effect of this was, of course, accentuated by the failure of the lights....

... all exits normally open to the public [soon] were useless. Pouring of fire [and compressed volumes of burning gases seeking outlets] through such exits made it impossible for humans to pass simultaneously through these exits safely. In the course of such pouring, the mass of burning gaseous material appears to have depressed from its high elevation within the premises in order to pass through the exits. The finding of bodies piled up at many of the exits is attributable to this....

... Some few persons passed through [the main revolving door off Piedmont Street] before the mass of flame actually reached it. The door appears then to have jammed. [However] there was a very great pouring of flame through this exit, [and] apart from jamming this

> door could not, by reason of such pouring of fire, have served as an available exit once the mass of fire and flaming gas had reached it....
>
> Persons unable to escape through the exit doors were thus exposed to the effects of the carbon monoxide gas, the superheated air, or the flames themselves....
>
> The death certificates signed by the Medical Examiner bear out these conclusions, as do the hospital records describing the appearance and condition of victims treated.

In estimating the extent of property damage, the report made this further point:

> The fire conditions alone, while fatal to many occupants, were at no time of sufficient size to challenge the resources of the Fire Dept. after response to the alarm.
>
> Rescue work was the first object of the responding fire companies. Had the building been unoccupied the fire could have been extinguished even more promptly than it was.... It was a quick-burning fire, which expended itself soon after the firemen attacked it; but certain portions of the building [the roof structure, for example] burned for a longer time than would have been the case under different conditions....

CAUSE OF THE FIRE.

From all the evidence before me [concluded Commissioner Reilly] I am unable to determine the original cause or causes of this fire.

I find no evidence of incendiarism.

A bus boy, aged sixteen... testified to lighting a match... and dropping the match to the floor and stepping upon it. After a careful study of all the evidence, and an analysis of all the facts presented before me, I am unable to find the conduct of this boy was the cause of the fire.

I have investigated and carefully considered, as possible causes of the fire, the following suggested possibilities: alcoholic fumes, inflammable insecticides, motion picture film scraps, electrical wiring, gasoline or fuel oil fumes, refrigerant gases, flame-proofing chemicals. There is no evidence before me to support a finding that any of these or any combination of them caused this fire.

This fire will be entered in the records of this department as being of unknown origin.

RECOMMENDATIONS.

1. Installation of automatic sprinklers in any room occupied as a restaurant, nightclub, or place of entertainment.
2. Prohibition of the use of basement rooms as places of assembly, unless provision is made for at least two direct means of access to the street, with installation of metal-covered automatic-closing fire doors required in any passage between basement room and first floor.
3. Requirement of defined aisle space between tables in restaurants, such tables to be firmly affixed to the floor to prevent upsetting and obstruction of means of egress.
4. Exit doors in places of assembly to have panic locks and no others. Such exits to be marked by illuminated "EXIT" signs (powered by a supplementary electrical system which would not be affected by a failure of the main lighting system).
5. Absolute prohibition in places of assembly of any fabric or material containing pyroxylin [a highly combustible cellulose nitrate mixture].
6. Absolute prohibition in such places of any suspended cloth false ceiling.
7. Window openings of sufficient area, equipped with louvers with a fusible link so as to open automatically

when subjected to heat, should be required in basement rooms used as places of assembly. *A major lesson of this fire is that persons and fire must be provided with separate means of exit....*

The report concluded with a complete, final list of all confirmed casualties of the Cocoanut Grove:

Injured (hospitalized only)—166.

Dead—490.

These recommendations and other appeals for tighter regulations by such organizations as the National Fire Protection Association gained rapid and wide support all across the United States and Canada. (In the weeks immediately following the Grove fire, New York City, then the nightclub capital of North America, reported record lows in patronage—down an average 50 to 75 percent.) As a direct result, stringent new laws were soon to be enacted by municipal and state governments around the country.

Thus, if calamity can ever truly be said to breed benefits, this one terrible experience at least had the effect of forcing antiquated safety codes into the twentieth century. As it had opened the eyes of medical science to more effective methods of treating fire victims.

Clifford Johnson felt there was just one more thing he needed to do before leaving Boston City Hospital—and this was *personal*. As he returned to nearly full health, he'd shown himself to be an outgoing, even cocky kid, an irrepressible flirt with the young nurses, and noticeably vain about his appearance. What bothered him the most, after all he'd gone through, was the bald patch at the crown of his head, where Dr. Browder had told him hair would never grow again. It became a private fixation with him, as though some mark of shame: twenty-two (now) and *bald*? Heck, no!

He wasn't about to leave the hospital looking like *that*.

So, in his final weeks there—moving about very well by himself and even permitted outside the hospital for walks—

unbeknownst to Browder or any of the others, he conspired with Ellie Kampper to make an appointment with a salon downtown where he could be fitted for a hairpiece. One day Ellie came in early and met him outside, and they took a cab to the hair stylist's.

He was fitted perfectly, and he returned to the hospital in smug triumph, hardly able to wait for the surprised comments and compliments. But nobody noticed—or at least nobody said anything. Dismayed, disgruntled even, Johnson at last confronted Dr. Browder and Mercy Smith. Didn't they notice anything different about him?

They looked at him blankly.

Johnson turned and pointed to the top of his head. "Look!" he demanded.

It took them a second or two to understand; they'd grown so used to the way he looked that they hadn't really seen the bare patch anymore. Then Browder exclaimed: "Oh! I see. Well, my goodness...."

"What is it?" the nurse asked dourly.

"It's *hair*! What do you think?"

The two were silent for a moment. "It's nice," she said. And that was it. They left him forlorn as a little boy whose attention-seeking mischief had gone unnoticed.

But it was Mercy Smith who combed his new hairpiece carefully into place and helped him put on his fresh set of Coast Guard blues for the first time in almost exactly a year when, on November 26, 1943, just two days before the anniversary of the great fire, Clifford Johnson was discharged from City Hospital. She took him downstairs, where photographers and reporters waited. Johnson stood lean but fit in his uniform. He smiled and thanked everyone and hugged his nurse for the cameras. Then he got into a Navy car and was driven away to the Brighton Marine Hospital, where he would continue his convalescence.

It was not an emotional farewell, because they all knew the young man would be back. Dr. Browder had some final repair work still to do on him.

* * *

Johnson remained in Boston another nine months, resident at the Marine Hospital but returning to City regularly for continued surgical treatment of his infected leg bone.

He was by now truly a scientific "marvel," the subject of thousands of pages of case reports and studies in medical journals all over the world. And there was more to come— volumes still to be written on the single most comprehensive, and astounding, history ever of individual burn damage and its successful treatment.

Newton Browder, at the center of the story, had come into quite some demand as a lecturer on the subject, with invitations coming from medical and lay groups throughout New England, the rest of the United States, and abroad. He had worked up a most dramatic presentation, combining his personal involvement with a stunning selection of slides made from the more than six hundred color photographs taken of Johnson almost from the time he'd been brought in.

For one such lecture before a seminar of student nurses at Newport, Rhode Island, in the late spring of 1944, Browder for the first time asked Johnson himself if he would appear as well. Johnson was hesitant—in the hospital he'd always felt uncomfortable being observed by white-gowned strangers as if he were some kind of guinea pig or freak—but Browder persuaded him they could make an outing of it: Mrs. Browder would come, and Leo Goodman, the medical photographer whom Johnson had come to know well, and they would spend the day at Newport Beach before the evening lecture. All in all it would be a refreshing break from the humdrum hospital routine. And then, of course, there would be all those young nurses in training.

The four picnicked on the beach with lobster and homemade potato salad and hot apple pie wrapped in newspapers. And then, content and relaxed, they all went to the lecture. Johnson in uniform, sat out front with Mrs. Browder and Goodman as the doctor, onstage, gripped the audience and held it enthralled with his now set piece, orchestrating the suspense by adroit interpolation of the starkly graphic slides. The tense silence was broken only by astonished gasps and little cries of sympathy. When he finished, the assembly of young women was utterly wrung out.

Then Dr. Browder asked that the house lights be turned up. He said he had with him someone they all might like to meet. He motioned Johnson up to the stage.

They stood side by side, and Browder, smiling broadly, announced: "Ladies, may I present Clifford Johnson."

The audience gaped in surprise at the glowing young sailor standing up there so tall and straight and *handsome*, smiling out at them shyly—like a young god or a movie star—and then they rose as one and burst into a spontaneous ovation that went on and on, applause and cheering with some whistles and girlish squeals thrown in.

It was perhaps the most gratifying moment in any young man's life. Or in any middle-aged doctor's. Josie Browder beamed up at the two of them until she couldn't see anymore through the tears. Around her, the young nurses in training were weeping for joy as well.

That September of 1944, Clifford Johnson, twenty-three, after not quite two and a half years in the Coast Guard—almost two years of which he'd been in hospital—received an honorable discharge. He was going home to Missouri.

This time at Boston City Hospital the goodbyes were touching and teary. They would never forget him, nor he them. He promised he would come back to visit them all. No one really believed that; but some way or another, he would always be with them—one of the few inspiring memories to outlast the lingering nightmare of the Cocoanut Grove.

chapter 19

The court-appointed receiver and trustee of the Cocoanut Grove, attorney Lee Friedman, through his counsel Frank Shapiro, had labored two years to keep afloat hopes that at least *some* reparations might be forthcoming to the injured and the families of the dead.

For Shapiro it had become an obsession. He had tirelessly pursued every possible legal means of attaching, and *securing*, the remaining assets of the Grove for distribution to the hundreds of rightful claimants, even to the point of collecting and disposing of personal articles—garments and jewelry and such—lost or abandoned by the fire victims. (Where families requested return of this or that item, of course, Shapiro obliged; yet for the most part he'd found that few cared to retrieve such grim, and for many devastating, momentoes. Some of these items, with the family's permission, he tried to sell. Yet, most of the victims' possessions wound up being given away to various charitable organizations.)

Survivors' claims as originally filed had amounted to some $8 million. Many of these were judged excessive, and the total had been scaled down to around $2.5 million. But the discouraging bottom line, by 1945, was that from all available sources there was only about $200,000 to be divided among the five hundred or so claimants. And even that sum, all of it, was tied up by the federal government in tax liens against the Welanskys. So, on paper at least, there appeared to be nothing.

The attorneys doggedly refused to give up, however, and Friedman and Shapiro went again and again to Washington to argue with, plead with, cajole the Treasury and Justice departments on behalf of their constituents. And midway through 1945, they finally came back to Boston with new hope that the government might be willing to "settle"—perhaps, at least, to reduce its tax claim against the Grove by as much as half. That would mean approximately $100,000 left for them to distribute.

Friedman advised all the claimants: If there were no more hitches, they might expect to receive about 5 cents of each dollar claimed.*

Among the capital assets acquired by the receivers had been the Cocoanut Grove property itself, once they had been able to persuade Jennie Welansky to relinquish title to what, after all, was no more now than a worthless pile of rubble. Early in 1945, the parcel was sold for $15,000 to a firm called Film Exchange Transfer Company (an interesting irony, in that some two decades earlier the building that was to become the ill-starred nightclub had been occupied by Paramount Pictures Film Exchange). The new owner, whose offices were down Piedmont Street from the site, planned to raze the property and construct additional offices and a garage there.

For over two and a half years after the fire, the blackened ruins of the Grove had remained untouched, except for occasional nighttime raids by youthful explorers or vandals. Broken-down walls and doorways and the demolished roof had been boarded up for safety's sake as well as to deter foraging, but periodically the fencing had to be reinforced after one intrusion or another. There was never a watchman,

---

* The case would take another two years to resolve. After payment of state taxes, court fees, and legal expenses, each claimant eventually received about $160.

however, nor, after the first weeks following the disaster, any particular surveillance by the police. It had become a no-man's land, and doubtless its debris had been picked over hundreds of times.

But then, in June 1945, only weeks before the Film Exchange Co. was to begin clearing the site, the police, during a routine inspection of the vacant property, made an astounding discovery: the Grove had been broken into again, but not by random scavengers. This time somebody found and looted, and left behind empty, a strongbox that had been hidden inside a downstairs wall of what had been the Melody Lounge!

It was a Mosler, steel, about a foot long and six inches deep, and not a trace was left of whatever it might have contained. The combination lock had been neatly, expertly, drilled out, as though the thief or thieves had not had the combination. But it seemed plain to the police that whoever it was had known just where to find the box.

It had been secreted behind an inside wall of the stairway. In the littered, dungeonlike darkness, the intruders had apparently gone straight to the spot and attacked just that one section of wall—for there were no other signs of damage—slashing through the leatherette outer covering (which, perhaps interestingly, was only slightly singed in that spot) and splintering the wooden inner framework to get to a concrete shelf on which the strongbox had been deposited. Having cleaned it out, they'd discarded the box among the rubble, leaving not a single fingerprint.

What had been in it? Who had concealed it there? Who else might have known of it? The authorities were disconcerted, not to say rather embarrassed, that in all the official investigations and careful sifting of the Grove wreckage after November 1942—by fire inspectors, police, insurance adjustors, lawyers—none had so much as suspected, much less detected, the existence of such a cache.

There was immediate speculation that it might have belonged to or been connected with the notorious King Solomon—that it contained illicit cash, or gems, or possibly even secret records that could be incriminating to former Solomon associates still active in the rackets. But while that basement space had been there in Solomon's time, part of the property then adjoining the Grove, it had not been acquired until some years later by Barney Welansky, and converted into the Melody Lounge only in 1938, more than five years after Solomon's death. It didn't seem likely that the area could have been completely redesigned and restructured around the hidden safe.

Naturally Welansky was questioned about it in prison. Morosely, he disclaimed any knowledge of it. Obviously sick and weary of his life, he really didn't care anymore.

Who else might have known? The architect, Bodenhorn? His construction people? The longtime maître d'hôtel Angelo Lippi? Mickey Alpert (or his original partner, Jacques Renard, who, plagued by both marital and gambling problems, had long since removed himself to Florida)? Barney Welansky's loyal brother, Jimmy? Some even wondered about the man who'd originally financed the building of the Grove, the swindler Jacob Berman. Most of the money he'd absconded with had never been recovered, after all. But then, Berman had been nowhere near Boston when the place was being built and opened, and right after that he'd gone to jail. In any case, all those who could be reached professed no knowledge of the mysterious box.

The most intriguing question was, why had the thieves waited so long to retrieve it? Had they simply waited until the Grove was all but forgotten and unguarded? Or were they spurred by the knowledge that the premises were about to be razed for new construction?

No one seemed to know, or at least no one was talking. It remained, remains, an unsolved mystery—the final enigmatic twist to an already bizarre story.

Not long after, in mid-August 1945—at about the time Japan was surrendering to the victorious Allies—demolition gangs descended on the weathered ruins between Piedmont and

Shawmut streets to begin clearing the site. Soon it would seem as if the Cocoanut Grove had never existed; indeed, the city had passed a statute that no public place in Boston ever again could be named "Cocoanut Grove."

But, like the war in a way, as much as people would have wished to forget it, the memory would not die—not for anyone who'd ever spent a glamorous, enjoyable evening there, or for the survivors of its spectacular finale, and certainly not for those whose loved ones had died there.

Today, more than forty years later, the Cocoanut Grove evokes both awe and sadness, even among those who had not been born at the time. Few, however, are aware of the legacy of good it has left.

In November 1945, almost exactly three years after the event, researchers at the Harvard Medical School finally pinpointed the source of the mysterious sweet-smelling toxic fumes that had snuffed out the lives of so many.

Laboratory tests (interrupted during the war) disclosed the presence of a substance called acrolein in the imitation leather ("leatherette") used so extensively in the Grove's decor and furnishings, which under great heat released deadly poisonous vapors. It was the first time acrolein—derived from the sweet, oily alcohol known as glycerol, which is formed by the decomposition of animal fats with alkalis or superheated steam—had been isolated as a component in leatherette. Used as a plasticizer in the cellulose nitrate from which the imitation leather was made, its function was to make the material soft and supple and to give it a sheen. Acrolein was (and is) an ingredient in tear gas.

The Boston medical examiner, Dr. Timothy Leary, who had followed the Harvard research closely, said he knew of no other fire disaster in which victims literally had been gassed to death.

"There is nothing in medical history," he said, "that quite corresponds to this."

EPILOGUE

Following is a colloquial adaptation (by the author) of a psycho-medical report on one victim of the Cocoanut Grove:

> A young man named Francis Gatturna and his wife, Grace, had been in the Grove's show room and were felled and separated soon after the fire broke out. He came to out on Shawmut Street and saw no sign of his wife. Taken to Massachusetts General Hospital, Gatturna was judged to have suffered relatively little damage: he was conscious, lucid, showing no signs of shock. There was a small first-degree burn on his face, another second-degree burn on a knee, both clean; inhalation burns were minor and respiration regular. He was put in a bed for precautionary observation and administered the antibiotic sulfadiazine.
>
> He inquired once about his wife, but at that point she had not been located, and he did not ask again—as if fearful to hear. After five days, he was told Grace had died. He actually seemed relieved to know one way or the other, and in the few more days Gatturna remained in the hospital he impressed the staff by his apparent self-control and improved spirits. Physically, he recovered fully and was discharged on December 8.
>
> However, on New Year's Day 1943, Gatturna was

returned to Mass General by his extremely concerned family: he was in a state of severe, possibly dangerous, mental depression.

At home he had turned unrelievedly restless, by turns going through phases or fits of agitation, preoccupation, brooding morbidity. It had become almost impossible to carry on a rational conversation with him because his mind would wander, blocking out awareness of others. Often he would suddenly complain of unbearable tension, of having trouble breathing, of a mounting fear that something terrible was going to happen, of being pushed by some unknown force toward an act of violence. He would mutter aloud to himself—always about the night of the fire, as though reliving it over and over: he had wanted to help Grace, *would* have, but he had fainted and lost her; and he was saved, but she had died; *he* should have died too. Lashing himself with increasing bitterness, he felt he must be doomed.

At first, psychiatrists at the hospital were unable to penetrate these overwhelming feelings of guilt and self-hatred. Gatturna was given large doses of sedatives to try to relax him, but even with them he slept so fitfully, crying out fiercely, that special nurses were assigned to stay with him night and day.

But then, about his fourth day back at Mass General, all at once he appeared to have come out of it. He was composed, less fevered, civil to those attending him, even receptive to the doctors' attempts to treat him. They felt they might start making some progress at last, that he had overcome or at least come to grips with his destructive obsession with guilt.

The patient continued in this passive, more amenable mode for several more days. Then, on January 7—having playfully distracted the attention of a nurse lulled by his improved humor—Francis Gatturna hurled himself through the closed window of his hospital room and in a splintering of glass plummeted several stories to his death. He was just thirty years old.

Attorney Frank Shapiro, who, as counsel to claimants against the Cocoanut Grove, had been over a period of time in uniquely intimate association with survivors and families of the dead, years later would say that he knew of "a number" of those victimized one way or the other by the disaster who were "still in mental institutions—hopelessly insane."

Shapiro, however, would not venture to be more specific (in observance of his obligation to preserve confidentiality), and this author has found no other documentation of his assertion—thus there might be some reasonable question as to its accuracy. Nonetheless, from examination of other published sources one may judge that what Shapiro says is not beyond credibility. For scientific observations in the years following the Cocoanut Grove did in fact discover a remarkably high incidence of residual psychological disorder growing out of the calamity.

Dr. Alexandra Adler, of the Harvard Medical School's Department of Neurology and the Neurological Unit at Boston City Hospital, reported in 1944 that 54 percent of Grove survivors interviewed were found afflicted with "post-traumatic neuroses"; and at Massachusetts General Hospital, 44 percent of the victims studied by Dr. Erich Lindemann, then a member of its Department of Psychiatry as well as being a neuropathologist at Harvard, had developed numerous "neuropsychiatric problems" after the fire. Moreover, Dr. Lindemann observed in a surprising number of the relatives and even friends of victims an "emotional upset [that] attained the proportions of a major psychiatric condition and needed trained intervention."

The symptoms of psychic damage were common, in varying degrees, to victims and their loved ones alike: grief, guilt, anxiety, depression, recurrent nightmares, either apathy or hyperactivity, withdrawal from or inability to resume normal activities, a sense of unreality, persistent resentment, and always the fear, reasoned or imagined, of a new disaster about to happen, fear of being in crowds, of entering a restaurant, of sitting in a theater. For many it would lead to lives of restlessness or inconsistent behavior, of indefinable dissatisfaction, even of complete incapacity.

Without belaboring the point, it would seem accurate to say that the effects of the Cocoanut Grove disaster reached far beyond the immediate event. The case of Francis Gatturna may have been extreme—it is not known if there were other suicides that could be so directly traceable to that fire—but there is little question that a great many people were left fearfully branded by more than physical injury.

So Frank Shapiro could have been right.

Here's how a few of the other lives affected by this drama turned out:

Joseph Buccigross. After eighteen months of suspension without pay, Buccigross's case was reviewed by Attorney General Robert Bushnell, who, in July 1944, went before the state's supreme judicial court to say that no true evidence of dereliction of duty could be found to warrant prosecution. On his recommendation, the case was dropped. Buccigross was restored to active duty, but remained stripped of his former command and relegated to a desk job at Boston Police Headquarters. (It would take a special bill by the state legislature to award him all eighteen months' back pay.) He stayed on the job—a pariah to some of his fellow officers, and still resented by many in Boston who had suffered from the fire—to mandatory retirement. And then he fought for, and won, a claim for disability benefits for the superficial injuries sustained while on duty that night at the Grove.

Barnet Welansky. In September 1943, Welansky's frail health had deteriorated and he was transferred from garment-making at Charlestown State Prison to lighter work at the minimum-security Norfolk Prison Colony. (On the basis of his health, his lawyer, Herbert Callahan, at that time petitioned for a reduction in his twelve-to-fifteen year sentence, but the plea was rejected.) In June 1944, Welansky's manslaughter conviction was upheld by the Supreme Court. In February 1946— almost three years after that conviction—he was formally disbarred by the state bar association.

By then his health had failed alarmingly; in June his ailment was diagnosed as cancer of the right lung and trachea or windpipe. He was granted leave from Norfolk that summer to enter Mass General for radiation treatments (X-rays having just recently come into use as a weapon against cancer). In September he was returned to the prison colony, his disease having been diagnosed as inoperable and almost surely terminal. Herbert Callahan appealed for a pardon directly to Governor Maurice Tobin—with whom Welansky allegedly had been so "tight" when Tobin was mayor of Boston.

In November, after much soul-searching (and indeed hard political deliberation), Governor Tobin announced Welansky's pardon. There was a storm of outraged protest, but the governor did not back off. On November 26, 1946, Welansky, shrunken and infirm, walked out of prison. To the reporters gathered outside, he made his first public statement since his trial: "I only wish I'd died then, with the others." A few months later, early in 1947, he did die at the age of fifty.

Daniel Weiss. Welansky's faithful nephew was accepted as an intern at Boston City Hospital in the spring of 1943. Nine months later he was drafted into the Army and, having been in the ROTC throughout medical school, was commissioned a first lieutenant in the Medical Corps. Trained in military neuropsychiatry, he went to England and France and was discharged a captain in 1946. After further psychiatric studies, Weiss joined the Veterans Administration as a staff psychiatrist in 1949, and by 1961 he was assistant chief of the Psychiatric Service of Boston's big VA hospital at Jamaica Plain. After that he went into private practice. In recent years he has also been chief of forensic medicine at Bournewood Hospital in Brookline, and has gained considerable recognition—and frequent exposure on the television news—as an expert forensic witness in major court cases.

Dr. Weiss has always stoutly defended his uncle. Barney Welansky was hard in some ways, a tough businessman, not a "personality," Weiss has conceded, but he was a man of his word, fair, and a literal disciple of the day's-pay-for-a-day's-work ethic. The doctor's most telling recollection of the kind of man his uncle was dates back to when Weiss, as a young medical student, was still new at his weekend job at the Grove. He was stationed at a side door early one evening before the club had opened, when a panhandler accosted him and asked for a dime. Weiss (who himself then earned from $2.50 to $3 a night) was fumbling in his pocket for some change when his uncle happened by. Barney grabbed his nephew's arm and said, "No, no. Watch me." Then he turned to the derelict: "This young fellow was going to *give* you some money. But that's no good. Instead, why don't you go down to the kitchen, tell the cook I've sent you—Barney Welansky, I own this place—and he'll give you a cup of coffee and then put you to work. At the end of the night, everybody who's worked gets a meal—and that includes you—and you'll get paid too. How does that sound?" The man blinked at him, let out a stream of obscenities, and shuffled down the street still muttering foully. Barney turned to young Weiss: "See? Remember this. Don't ever *help* anybody be a bum!"

Angelo Lippi. The noted maître d'hôtel, "the Count," whose name and presence from the beginning were virtually synonymous with the Cocoanut Grove—and who ironically was spared its inglorious end by a case of gout—recovered his health and went on to supervise many other restaurants and clubs. He died in 1966, at the age of eighty.

Mickey Alpert. The Grove's MC moved to New York with Kathryn Hayman early in 1943, where for several months he recuperated at the Riverside Drive apartment she shared with her mother. That summer, having shaken off his depression, he accepted an engagement in the Catskills as MC and orchestra leader. In the fall Mickey and Kathryn were married— at the Central Park West apartment of their friends Milton and Joyce Berle. He became a fixture at the Rio Bamba nightclub in Manhattan. When they had a child, a daughter, in 1946, Kathryn gave up dancing professionally and they settled in suburban New Rochelle (where for a number of years Kathryn operated a successful dance studio on Main Street).

After World War II, Mickey also eased out of the fading nightclub business and went to work for the Kudner advertising agency, which had with great foresight committed itself heavily in the exciting new medium, television. In those days, enterprising ad agencies produced a good portion of regular TV programming, and Mickey became the casting director for many of the top variety shows, including Milton Berle's "Texaco Star Theater" and the early Jackie Gleason comedy revues.

Over the next fifteen years, Mickey made himself a power not only at Kudner but also in network TV and a well-known, popular figure in the entire New York entertainment scene. (He was a regular at the venerable performers' club, the Lambs, where he became practically the "house" MC for its various professional affairs.) But television management had been changing during those years, and the agencies were being stripped of their control of programming; and when, by the start of the 1960s, Kudner lost some of its most lucrative accounts and had to cut back its costly TV involvements, Mickey left to set up his own booking office.

It did not fare well ("He *gave* jobs away" to the show-business people, big and unknown, he loved so well, his wife Kathryn would complain), and though he never stopped hustling, he lost a lot of money. He also developed a bad case of high blood pressure—so bad, finally, that he was refused life insurance because of it.

One night in September 1965, as he did so often, Mickey attended a testimonial dinner at the Lambs for which he'd organized the entertainment. He was not feeling his usual ebullient self, and afterward he went uptown for a nightcap at Toots Shor's restaurant, a celebrity hangout which was one of his haunts. Then he collapsed. Rushed to the Medical Arts Hospital on 57th Street, he died within an hour of a massive stroke. Mickey was sixty-two. (He was still survived, in 1983, by his widow and daughter.)

George Alpert. Mickey's eldest brother—the patriarchal protector—would achieve eminence as a corporate attorney. Among other endeavors of note, in the 1950s he took on the presidency of the venerable and sorely troubled New York, New Haven & Hartford Railroad, which was badly in need of reorganization. But after several years at what proved to be an "impossible" task, he returned to the law. Alpert retired from active practice only in 1983, at age eighty-four.

Dorothy Myles...Al Willet (and *Pepper Russell*)...*Jack Lesberg.* Dottie Myles underwent hundreds of skin grafts to her disfigured face and hands and was under special care for nearly a full year after the fire. But with great spirit and determination—she overcame the cardiac weakness that had complicated her recovery—and the skill of the various surgeons attending her, she gradually recovered and her appearance was restored almost to its original loveliness. She relearned the piano and singing and after several years resumed performing. Never the "star" many had thought she might be, she was nevertheless able to work regularly in some of the better cafés and lounges in the Northeast. (Al Willet ran into Dottie—to his amazement and delight—in New York later in the 1940s. At that time, so soon after the fire, she still wore gloves and veiled hats. But in later years she became more confident and no longer masked herself.) A New York newspaper interviewer described her in the 1960s, when she was appearing at the Henry Hudson Hotel, as "a beautiful woman with red hair and black eyes."

Al Willet and Jack Lesberg, the Grove musician pals,

were in Boston City Hospital about the same length of time—into early 1943. Both had lost their most valuable possessions, their instruments; but upon their recovery the Red Cross provided them each with replacements—Willet a saxophone and clarinet, Lesberg a bass. It was six months before Willet's lungs and throat were healed enough for him to be able to play. In 1944 he joined the Shep Fields orchestra and went with them to New York, where they were booked into the Copacabana. Next he was with the Jerry Wald band. Still later, Willet switched to personal management, first for actor Roddy McDowall, and then after moving to California, for orchestra leader and radio personality Skinnay Ennis, whom he managed until Ennis's death. Willet finally went into the lucrative mobile home business in California, and a few years ago he retired, a recent widower, to live alone in Hollywood.

Pepper Russell and Willet continued their romantic relationship for several years, even after he had moved to New York. But he did not want to remarry yet, as she did, and the romance died when Willet relocated to California—where, eventually, he did remarry.

For months after the fire, Pepper had not been able to bring herself to dance in a club, and she took a clerking job in a Boston Western Union office. But in time her nightmares faded and she resumed dancing—her first booking at Jimmy Welansky's Rio Casino. Later, she joined the national company of *Water Follies*, starring Buster Crabbe, and wound up in Hollywood, where she danced in a number of short-subject films. Returning to the East, she was booked into various clubs as a featured tap dancer until, in her thirties and tiring of traveling the club circuit, she opened a studio—Henrietta's School for Dancing—in East Braintree, near Boston. She has since remarried twice, and is now "retired" and living comfortably in an affluent Boston suburb.

Jack Lesberg returned to work at Boston's Hotel Statler a few months after the fire. Like Willet, he, too, moved to New York in the mid-1940s, and he played at Eddie Condon's then new jazz club in Greenwich Village. Lesberg—who plays viola, violin, and piano as well as bass fiddle—at the same time was with Leonard Bernstein's City Center orchestra, then with the NBC Orchestra. He played for radio's "Hit Parade" and "The Telephone Hour" for eight years, worked a number of early television programs, and did a lot of TV commercials. In recent years he has toured the world with both jazz *and* symphonic ensembles. Divorced and childless, Lesberg, when not traveling, keeps a bachelor apartment in New York City.

Martin Sheridan. In May 1943, two months after leaving Mass General, Sheridan was hired as a full-time reporter by the Boston *Globe*. Subsequently he went to Greenland and to North Africa as a war correspondent, then, in June 1944, to the Pacific theater.

One evening he was standing on the afterdeck of the U.S.S. *Fremont*, flagship of an amphibious force steaming toward the Caroline Islands, awaiting the start of the nightly movie, when a young seaman approached, studied him questioningly, then asked: "Excuse me, sir. Are you Marty Sheridan?"

Sheridan was surprised. "I am. Do I know you?"

"Probably not." The sailor smiled. "But I'm the one who took you out of the Cocoanut Grove the night of the fire."

Sheridan could only gape at him. He'd never known just who had dragged him from the fire—he'd always assumed it had been a fireman. And now, almost two years later, in the middle of the Pacific, halfway around the world—! The newsman hugged the youngster and cried, "Oh, God, thank you! Thank you!"

The sailor—Electrician's Mate First Class Howard Sotherden, from Rhode Island—explained how it had happened. He'd been on leave that Saturday night, "bouncing" in downtown Boston, when all the sirens had started. He'd run to the Grove, fought his way inside, and helped the firefighters remove several bodies. Back in the blazing club, near the Terrace in the show room, he stumbled over another fallen victim who seemed half-conscious. Pulling the man to his feet, he manhandled him out to the street. The man's face was smudged with soot, his hair singed, his eyes puffed, broken eyeglasses hanging from one ear. As the sailor laid him down, he heard one bystander exclaim to another: "Hey, that's Marty Sheridan!" (They were a couple of local newsmen, Sheridan later learned.) When Sheridan was taken off in an ambulance, young Sotherden had never expected to see him again.

Sheridan wrote for the *Globe* until 1951, when he left to go into public relations with the PR whiz Steve Hannegan. He remarried and for many years has worked and lived near Chicago.

Stanley Tomaszewski. Perhaps no one was so cursed by the Cocoanut Grove, or has borne the stigma so long—or with such dignity—as "the kid who lit the match that killed all those people."

Tomaszewski left Boston in May 1944 for the Army Air Corps and did not return for six years. Washed out of the pilot program because of a childhood neck impairment, he nevertheless was recommended for Officer Candidate School and, just after the war's end, was sent to Japan as a second lieutenant with the army of occupation. There he was assigned to the Inspector General's Headquarters in Tokyo as an auditor (having had a year's training in accounting at Boston College). Promoted to first lieutenant, and enjoying his job and Japan, he signed up for another hitch.

He also met a Kansas girl there, Betty, a civilian stenographer at General Douglas MacArthur's HQ, and they were married in 1947. In three years they had two children, a girl and a boy, and they were content. But as his second enlistment ran out, and Tomaszewski found he could not get a Regular Army commission because he lacked a college degree, they decided to return home so that he could complete his schooling. So, in the spring of 1950, Stanley Tomaszewski went back to Boston, still only twenty-four but now with a wife and a couple of kids.

The four of them moved in with his aging parents in their cramped, threadbare Dorchester flat, while Stanley resumed his studies at Boston College. (The remaining three years of his scholarship had been kept open for him; but as he was then eligible for a full ride under the GI Bill, which provided modest income to boot, he suggested that the university pass the grant on to some other deserving young fellow.) He worked steadily at various jobs to keep the family going while he was in school—and Betty, though primarily occupied with the children, also took part-time work as a department-store salesperson. Eventually, they were able to save enough to move the family out of the tenement apartment into a small house.

In 1953, at age twenty-seven, Tomaszewski received his degree in Business Administration from BC and immediately got a position as an auditor in the Boston office of a national accounting firm. After four years there he was hired away by an auditing branch of the federal government charged with overseeing the accounts of civilian manufacturers with major defense contracts.

For the next twenty-five years Tomaszewski worked exclusively for the government, bulldogging—and becoming widely known and respected by—the many high-technology companies doing defense work in the Boston area. He'd reached the GS-12 grade by 1982, when, at fifty-six, he retired to become a private consultant. (A few years earlier he'd also retired from the Army Reserve, a full colonel and deputy commander of the 76th Maneuver Training Command based at Warwick, Rhode Island, directing tactical deployment and evacuation exercises.)

Through all of this more or less typical middle-American "success" story, however, Tomaszewski had seldom been allowed to forget that to some people he was still, would always be, a murderer.

It had begun again almost as soon as he and his family returned from Japan: the harassing, accusatory, often vile anonymous phone call, usually in the middle of the night; the occasional oath hurled by a passerby in the street; the incredulous-turning-scornful reaction of some people to whom he was introduced at business or social gatherings, the abuse and hurt endured by his wife and bewildered children.

It would diminish gradually as the years passed; but it would never end. In November 1970, twenty-eight years after the Cocoanut Grove, the Boston Fire Department—under a new commissioner, James Kelly—officially closed the file on the fire by issuing a "final" report on its investigations. The report was substantially a reprise of the earlier findings: *After a careful study . . . unable to find that the conduct of the (bus) boy started the fire. No proof of incendiarism . . . the Department is unable to determine the original cause or causes of the fire.*

Yet even following this renewed exoneration of Stanley Tomaszewski (whose name was not used in the report), there came a new spate of angry phone calls to his home—always, it seemed, in the dead of night: "Liar! *Killer!* You should pay!" Often the caller would sound addled with alcohol, sometimes just maudlin with unsatisfied grief. Tomaszewski would haul himself out of bed and listen to their rantings, but there was little he could say. He would hang up wearily and reassure Betty or the kids (three by then) that it was all right, just another of those calls, it didn't mean anything. But it was a nightmare.

The ultimate absurdity, perhaps, is that lawyers contact him still, asking him to appear as an "expert witness" in negligence suits arising out of fires. In 1977, after a fire destroyed a crowded night spot in Kentucky, killing 164 young people, they offered to fly him down there to testify *from his own experience*—for the defense, or the prosecution, or maybe both, he was never sure which. That made him sicker at heart than he'd felt in years: that he should have come to be regarded as an *authority* on deadly fires because once, another world ago, he'd had the gross misfortune to have been in one. Did they expect him to explain how to *start* such a fire? Well, *he did not know!*

Will Stanley Tomaszewski pay for the Cocoanut Grove as long as he lives? Maybe. Almost every November, at each succeeding anniversary of the disaster, newspaper and wire service feature writers retell the horrors of that unforgettable night in 1942, and always a dramatic ingredient in those stories is the incident of the bus boy who struck the match. (And every so often an itinerant author will come along to dramatize it at even greater length.) And so the name of Stanley Tomaszewski continues to be remembered.

Dr. Oliver Cope. At the time of the Cocoanut Grove fire, Dr. Cope, then turned forty, had applied for an Army commission and had already been confirmed as a lieutenant colonel in the Medical Corps. With the extended pressure of that crisis, however, and the acknowledged success of his new method of burn therapy—which required him to supervise the writing of the many procedural and analytical papers required for national distribution—plus the fact that so many of Mass General's senior and staff physicians already had left for military duty, the hospital asked him to stay on in an administrative and teaching capacity. (Cope, a Quaker, was almost relieved.) For the rest of the war, he managed the hospital's growing burn care unit and continued to serve on the National Research Council's burns committee.

Twenty years later, as a direct outgrowth of the burn treatment inaugurated with the Grove emergency, Mass General was designated as the site and given the supervisory management of the world-renowned Shriners Burn Institute. Cope was its first director.

Primarily a surgeon as well as an endocrinologist, Cope's expertise and greatest interest lay in thyroid and parathyroid surgery. He was acting chief of general surgical services until his retirement in 1969 at the age of sixty-seven. After retirement he remained active as a senior surgeon at Mass General, as Professor of Surgery Emeritus at Harvard Medical School, a writer and a lecturer. In 1983, at the age of eighty-one, he was still going daily to his office at the hospital.

Dr. Newton Browder. After the fire, Browder wrote a number of valuable medical papers on burn treatment, particularly on the extraordinary case of Clifford Johnson. Browder remained a surgeon at Boston City Hospital for sixteen more years and continued teaching general surgery at the area's three main medical schools, Harvard, Tufts, and Boston University. For some years he was also the physician for Harvard's varsity athletic teams. At the time of his retirement in 1959, he was City Hospital's chief consultant in orthopedics. He kept his small office, with a roster of regular patients, after that; but, according to his wife, Josie, he was "lost" without the hospital and his beloved teaching to energize him. Ten years after retirement, Browder died of heart failure at age seventy-five.

[Since 1942, medical science has taken tremendous strides—growing directly out of the Cocoanut Grove experience—in improving deep-burn treatment. Where four decades ago a victim with more than 20 percent of the body surface destroyed was considered all but unsalvageable (which was, of course, what made the Clifford Johnson case, for one, so extraordinary), today a person who has sustained as much as 80 percent, sometimes up to 90 percent, full thickness burns may have a good chance not only of surviving but of being restored to active life and at least presentable appearance.

[To a significant degree, such a leap forward is directly traceable to the pioneer work performed separately in the Grove emergency by Dr. Oliver Cope and Dr. Newton Browder. On the one hand, a vital part of modern burn therapy is priority attention to replenishment and maintenance of vital body fluids to avert fatal shock, as championed by Cope; on the other, perhaps the most significant new development in protecting the severely burned patient against invasive infection, as well as in effective surface restoration, is a technique devised and being continuously refined by medical researchers of the Shriners Burn Institute at Massachusetts General Hospital: the early implantation of "artificial skin." (This latter advance is in a way ironic, in that it might be described as an enlightened, highly sophisticated variation on the primitive old procedure generally employed before 1942, and denounced by Dr. Cope, of "tanning" burned surfaces to form a protective shield of "skin.")

[The artificial skin—developed by Dr. John Burke, who succeeded Cope as head of the Shriners Burn Institute, with Prof. I.V. Yannas of the Massachusetts Institute of Technology—is implanted into and over deep burn wounds as soon as the destroyed tissue can be removed, thus closing the wounds almost at once to both infection and fluid loss. The artificial skin consists of a two-layer membrane: the dermal or underlying layer, which is permanent, is made of cowhide collagen, a protein that is the chief constituent of human connective tissues, combined with a chemical substance obtained from shark cartilage called chondroitin 6-sulfate; the epidermis or outer layer of the membrane is Silastic, a form of rubber-like plastic. While the dermal implant in effect coordinates and eventually becomes one with the live tissue underneath to form a whole new tissue, the outer covering is a temporary graft which in time, usually after four weeks or so, is removed and replaced by healthy epidermis taken from elsewhere on the recovering patient's body.

[The artificial skin, which is simple to manufacture, and which can be stored for long periods at room temperature, so closely resembles human skin that eventually it is almost indistinguishable from the real thing. But its more essential advantage is that because it has been found to be readily accepted by the body there is no need, as in the past, to administer immunosuppressive drugs to prevent rejection of "normal" skin grafts—a process formerly necessary but often dangerous, because such drugs tend to increase the risk of infection.]

Clifford Johnson. Nobody at Boston City Hospital ever expected to see the stalwart young Missourian again after he'd finally gone home in September 1944—nor, doubtless, had he himself ever really anticipated keeping his casual promise to return one day. But he did come back, early in 1946, and again as a patient.

That winter, working on the family farm, he'd reinjured the leg in which the osteomyelitis had given Dr. Browder such difficulty, and the bone infection flared up again. Veterans Administration doctors in Missouri had various suggestions as to where he should go for new surgical repair, but to Johnson there could be only one place: Boston City Hospital and Dr. Browder.

He was welcomed as though it were the homecoming of a native son who'd made good. All who'd nurtured him through his time of hell flocked about him: Dr. Browder, so pleased and proud of his handiworks; Mrs. Browder; the nurses Mercy Smith and Ellie Kampper (neither of whom worked regularly at City but who came from their homes to see him); Philip Butler, just returned from active naval duty in the Pacific, where, as a lieutenant commander at Guam, Iwo Jima, and Okinawa, he'd been decorated for bravery; and all the other staff doctors, nurses, and therapists still there who'd attended him at one point or another. He also was the object of much curiosity and attention from the newer members of the hospital team who'd only heard about this medical "legend." Not the least welcome of these, to Johnson's twinkling eyes, were the awed young student nurses who fluttered about him.

Johnson was at City almost four months—though ambulatory much of the time, and, of course, not in the grave danger he'd faced before—as Browder worked carefully to mend the diseased leg. And in that time one student nurse (whose daily task was to inject him with penicillin, which by then had come into general use) caught his roving eye and held it... and as the weeks and months went by, they fell in love. Her name was Marion Donovan, a pert, sparkling girl from Dorchester, who was in her final phase of nurse's training, scheduled to be graduated that September. In July, when he was ready to return home, Clifford asked her to marry him. She said she would, but not before September. He said he would wait for her right there—he wouldn't go home without her.

They were married in Boston in September, had a farewell party at City Hospital with all their friends, and headed for Missouri to begin a new life.

Sumner, where he'd grown up, was little more than a farming waystop—a co-op grain elevator, a gas station with attached café, a tavern, general store, post office—and out of deference to his city-bred bride, Johnson tried to make his livelihood in Kansas City instead. They ran a small restaurant for a while, but that didn't work out. He drove a cab while Marion did some practical nursing. But it was hard to get the country out of the boy, and he grew increasingly discontented with urban life. She saw it, naturally, and when he excitedly told her of a job opening as game warden at a wild-goose reserve only a few miles from Sumner, she agreed to return with him to his roots.

The Swan Lake Reserve was operated by the Missouri Conservation Commission, which recently had opened its eleven thousand acres for goose hunting the last two months of each year. The warden's job was made to order for Johnson—with his love of the outdoors, the freedom and serenity of nature, the land, animals. Marion could see the change in him, the contentment and renewed zest for life; he became active in a nearby American Legion post and Masonic order. As long as he seemed so happy, she resolved to make the best of the bucolic life.

They'd corresponded regularly with the people back in Boston, sending letters and holiday greetings a few times each year, and in December 1956 they'd already sent off their Christmas cards.

One snowy afternoon just before Christmas, Johnson was driving home from the reserve through thick ground mist in his warden's Land Rover. He missed a turn, spun the wheel, the vehicle glanced off a snowbank and flipped over into a ditch, pinning him beneath. Gasoline from the ruptured fuel tank sprayed over the hot engine and set it aflame.

A farmer creeping past in a pickup truck stopped and frantically tried to pull the trapped, screaming driver free, but he could not get close enough to the blazing hulk. He raced off to get help, including the one nurse in the area.

By the time the nurse, Marion Johnson, got there—not knowing who the victim was—Clifford Johnson had been incinerated.

Fate, once denied, had been unforgiving.

It would be stretching dramatic license to say Clifford Johnson was the last victim of the Cocoanut Grove. But would not his life, and his end, have been different but for that one night on the town in Boston, November 28, 1942?

Postscript.

In New York, in December 1975, a fire at the Blue Angel nightclub on East 54th Street took seven lives. The deaths were attributed to a lack or inadequacy of automatic sprinklers, fire alarms, and emergency lighting.

In the shocked aftermath, the city enacted a new fire-safety law covering nightclubs, which went into effect on January 1, 1980.

Three years later, in mid-1983, the New York Fire Department found that more than *two-thirds* of the city's 687 licensed nightclubs still had yet to comply with—or were flagrantly defying—those safety regulations.

EDWARD KEYES is a native New Yorker who, over the years, has spent a good deal of time in Boston. A former journalist, he is the author of *The Michigan Murders*, which won an Edgar award for nonfiction suspense, and of the suspense novel *Double Dare*. He also collaborated with Robin Moore on the award-winning *The French Connection*. Mr. Keyes lives in Westchester County, New York.